Clinical Use of Anti-infective Agents

Robert W. Finberg · Roy Guharoy

Clinical Use of Anti-infective Agents

A Guide on How to Prescribe Drugs Used to Treat Infections

Robert W. Finberg
Richard Haidack Professor and Chair
Department of Medicine
University of Massachusetts
Medical School
Worcester, MA, USA
Robert.Finberg@umassmed.edu

Roy Guharoy
Chief Pharmacy Officer
UMass Memorial Health Care
Clinical Professor of Medicine
University of Massachusetts
Medical School
Worcester, MA, USA
Roy.Guharoy@umassmemorial.org

ISBN 978-1-4614-1067-6 e-ISBN 978-1-4614-1068-3
DOI 10.1007/978-1-4614-1068-3
Springer New York Dordrecht Heidelberg London

Library of Congress Control Number: 2011942229

Printed on acid-free paper

Springer is part of Springer Science+Business Media (www.springer.com)

Contents

Part I
Introduction to Anti-infective Therapy, Structures and Mechanisms of Action of Commonly Used Antibacterial Agents

Chapter 1
Introduction and History

The origin of anti-infective therapy in the Western hemisphere dates back to the use of cinchona bark to treat malaria by the indigenous peoples in South America. This discovery was taken to Europe in the seventeenth century. These observations were used to develop the drug quinine, which is still used in the treatment of malaria today. The Chinese have a 5,000 year history of using herbal remedies that may still lead to new anti-infective agents.

The first commercially available anti-infective agents were developed by German chemists who screened dyes for their anti-bacterial activities by injecting them into mice challenged with streptococci. This in vivo challenge resulted in the discovery of the compound Prontosil, which was marketed not only for treatment of streptococci, but also staphylococci and a variety of different gram-positive and gram-negative organisms. Interestingly, in terms of the later history of anti-infective discovery, Prontosil had no activity when used in vitro. The active compound, sulfanilamide (see Chap. 3), required "bioactivation" by the host and was a breakdown product of the original compound.

Prontosil was widely marketed in the US and other areas in the 1930s and used to treat a variety of diseases. This drug is not a natural product and therefore is not an antibiotic but an anti-bacterial agent. The first antibiotics were isolated from bacteria. The agent streptomycin was first isolated from bacteria in the soil in the laboratory of Selman Waksman, who originated the term "antibiotic." Waksman described an "antibiotic" (an odd term suggesting anti-life) as anything produced by a microorganism that inhibits another microorganism in high dilution. Streptomycin was marketed in the 1940s after a randomized clinical trial (one of the first in history) demonstrated it to be effective in the treatment of tuberculosis. It is the original aminoglycoside antibiotic. Like many antibiotics it was originally isolated from a species of *Streptomyces*, hence the suffix "mycin."

Penicillin, isolated after an accidental observation based on the ability of fungal contaminants to inhibit the growth of bacteria, was the first "miracle drug." It was isolated from the *Penicillium* fungus. The penicillins, cephalosporins, aminoglycosides, and carbapenems were all natural products isolated from microorganisms,

R.W. Finberg and R. Guharoy, *Clinical Use of Anti-infective Agents: A Guide on How to Prescribe Drugs Used to Treat Infections*,
DOI 10.1007/978-1-4614-1068-3_1,

and therefore are properly termed "antibiotics." The sulfonamides, the quinolones, and the oxazolidinones are all chemically synthesized and therefore are not antibiotics, but anti-infective agents.

Today antibiotics continue to be isolated from the products of bacteria and fungi. In the last two decades, the concept of "rational drug design" has led to the development of drugs based on differences between mammalian and fungal cells. Many of the anti-fungal agents, such as the echinocandins, are based on their ability to inhibit specific enzymes found in fungi but not in animals. Similarly, anti-viral agents such as acyclovir take advantage of enzymes that are important to viruses but not to animal cells.

Chapter 2
Basic Principles of Drug Delivery and Dosing

Mechanism of Actions of Anti-bacterial Agents

Anti-bacterial drugs work by exploiting differences between mammalian and microbial physiologic processes. Since bacteria, unlike mammals, have cell walls, many anti-bacterial agents work by binding to components of the cell wall or inhibiting the synthesis of the cell wall. These agents include the beta-lactam and glycopeptide antibiotics. Other anti-bacterial agents function by exploiting differences in DNA replication (the quinolones inhibit DNA gyrases that are used in bacterial replication) or transcription (rifampicin). Still other agents exploit differences in the translational process, as bacteria use ribosomes that are different from mammalian ribosomes. The macrolides, tetracyclines, and aminoglycoside antibiotics exert their anti-bacterial effects by binding to different components of the bacterial ribosome. Still other anti-microbial agents such as the sulfonamides function by exploiting differences in the metabolic pathways between mammals and bacteria. Since bacteria are unable to use folic acid but must synthesize it from para-aminobenzoic acid, drugs that inhibit this pathway will selectively inhibit bacterial growth without toxicity to mammalian cells (see Fig. 2.1).

Role of Pharmacokinetics and Pharmacodynamics in the Management of Bacterial Infectious Diseases

The minimal inhibitory concentration (MIC) for anti-microbial agents is the concentration at which the drug inhibits growth of the organism in vitro. Historically, the doses of anti-bacterial agents for use in humans have been chosen based on the MIC of the drug for the specific pathogen being treated. Drug levels are typically measured in human serum. However, dosing based solely on MIC may not be appropriate since MICs of certain pathogens labeled as "susceptible" can be higher than peak serum concentrations of some of the anti-microbial agents. Factors such as

R.W. Finberg and R. Guharoy, *Clinical Use of Anti-infective Agents: A Guide on How to Prescribe Drugs Used to Treat Infections*,
DOI 10.1007/978-1-4614-1068-3_2, © Springer Science+Business Media, LLC 2012

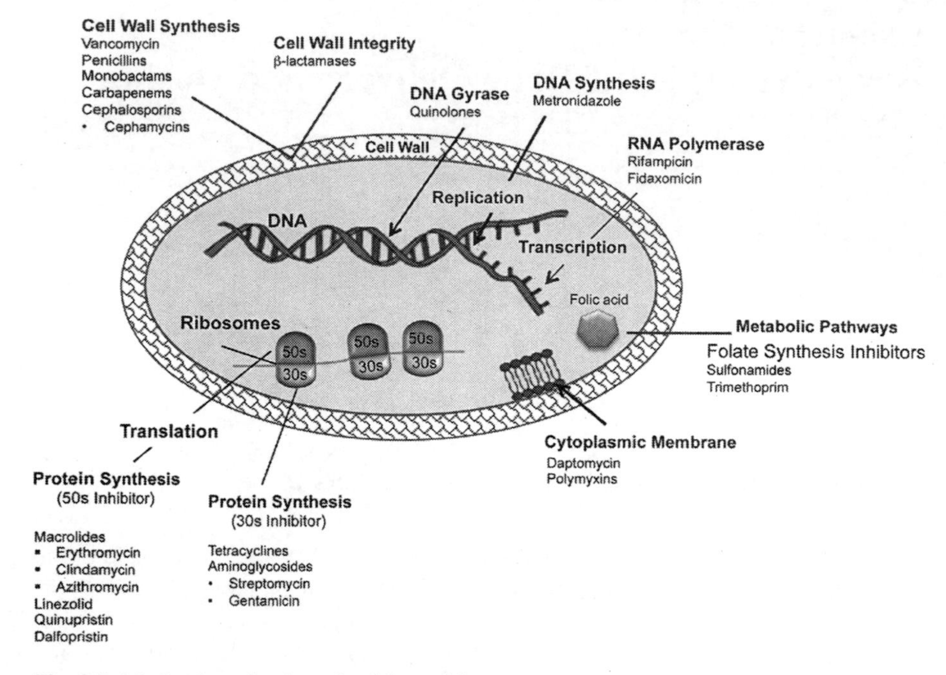

Fig. 2.1 Mechanism of action of anti-bacterial agents

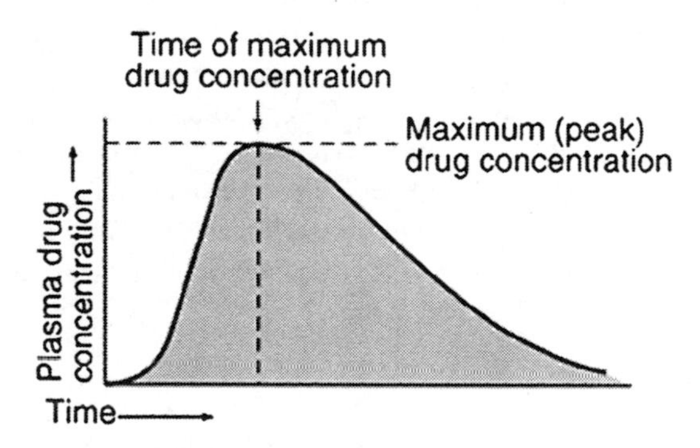

Fig. 2.2 Peak serum concentration over time

specific anti-microbial agent, pathogen, and individual patient play an important role in achieving the desired outcome. Most importantly, inadequate serum concentrations at the site of infections may lead to anti-microbial resistance (Fig. 2.2).

Pharmacokinetic (PK) and pharmacodynamic (PD) principles are the tools used to optimize anti-microbial therapy for individual patients. PK and PD studies help us to understand factors such as the onset, magnitude, and duration of drug response that allow for optimization of specific therapies. PK and PD determine how much

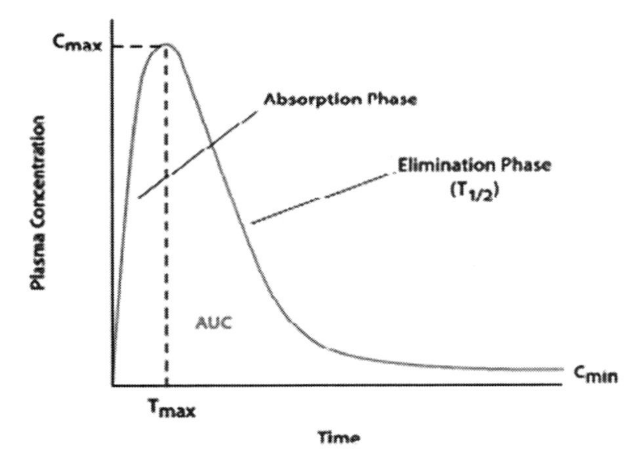

Fig. 2.3 Plasma concentration–time curve (*AUC* = area under the serum concentration time curve = a measure of plasma concentration of a drug over a time period and is derived from the area under the plasma drug-concentration–time curve). It is estimated by measurement of drug concentrations at various times; C_{max} = highest plasma concentration observed; C_{min} = lowest plasma concentration observed; T_{max} = time at which highest C_{max} occurs after administration of an extravascular dose; $T_{1/2}$ = time for a given concentration to reach 50% determined by volume of distribution and clearance; clearance = measure of efficiency of removal of a drug from plasma and it is a ratio of dose administered to AUC; volume of distribution = volume of body fluids into which a drug distributes at equilibrium

and how often a drug should be administered. One can think of PK as "what the body does to the drug" and PD as "what the drug does to the body." The integration of PK/PD helps to assess the interactions between a pathogen, host, and anti-microbial agent.

The pharmacokinetic (PK) profile of a drug describes its absorption, distribution, metabolism, and elimination. The PK defines the time course of drug concentrations in the body, tissues, and fluid. In essence, PK tells us how drug concentrations in the body change over time after administration of a dose. The parameters involved include: bioavailability (proportion of drug absorbed into the systemic circulation after drug administration; intravenous drugs are 100% bioavailable and other forms are mostly less bioavailable); minimum serum concentration of drugs (C_{min}); peak serum concentration of a drug following administration of a dose (C_{max}); time to peak serum concentration (T_{max}); volume of distribution (Vd – a relative measure of distribution of a drug throughout the body); area under serum concentration–time curve (AUC); elimination half-life ($T_{1/2}$ – time required for serum concentration reduced by 50%); and amount of time serum concentration above the minimum inhibitory concentration (T > MIC) (see Fig. 2.3).

The pharmacodynamic profile (PD) describes the impact of anti-microbial plasma concentration and response. It describes the effects of an agent on human cells and the pathogen. Interestingly, the body of knowledge in PD is limited

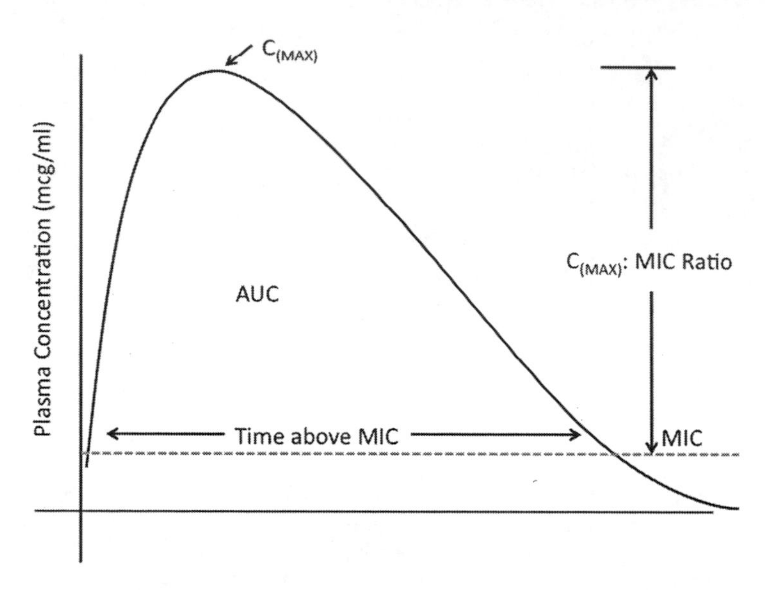

Fig. 2.4 Plasma concentration over time depicting pharmacokinetic–pharmacodynamic relationships. Activity of a concentration-dependent antibiotic can be predicted by AUC/MIC or C_{max}/MIC ratio requiring administration of adequate dose to achieve high concentration; on the other hand time-dependent agents require frequent administration to maximize T/MIC

relative to the PK studies, which can easily be determined. Drug concentrations in plasma are monitored at different times which can be used to determine the primary PK parameters such as clearance and volume of distribution. On the other hand, the study of PD is difficult since it is hard to determine objective, precise ways to quantify drug response. PD parameters include the time that a drug's serum concentration remains above the MIC for a dosing period (T>MIC), ratio of the maximum serum antibiotic concentration to MIC (C_{max}/MIC) and the area of the concentration time curve during a 24-h time period (AUC) divided by the MIC (AUC/MIC). Some agents continue to exhibit bactericidal effects even after clearance of the agent from the infected site. These post-antibiotic effects (PAE) have been observed with inhibitors of nucleic acid and protein synthesis (including aminoglycosides – see Chap. 9, and quinolones – see Chap. 12) and with beta-lactams against *Staphylococcus aureus*. Nucleic acid and protein synthesis inhibitor agents with significant PAE prevent pathogen growth even after serum concentrations fall below MIC.

Combined use of PK and PD helps us to optimize effective use of an antimicrobial agent. Integrated use of PK and PD data provides a rational basis to understand the impact of various dosage regimens on the time course of pharmacologic responses. It provides information on the effective dose and duration of therapy of a specific agent against a specific pathogen (Fig. 2.4).

PK/PD Profiles of Anti-microbial Agents

Time-Dependent Agents (Duration of Exposure to a Specific MIC)

For this class of agents, anti-microbial activity occurs after reaching the maximum threshold and then stops after the serum concentrations fall below the MIC. Bacterial killing is achieved once a threshold is reached. T>MIC results in anti-bacterial activity and is expressed as a percentage of dosing interval. The time above MIC can be extended by frequent dosing.

Beta-lactams (penicillins, cephalosporins and carbapenems) reach peak efficacy at concentrations approximately 4–5 times above the MIC and further higher concentrations do not result in increased bactericidal activity. The magnitude of pathogen killing is determined by the duration of exposure of the bacterium to the drug (T>MIC) rather than the degree of exposure above the MIC. T>MIC between 40 and 70% of the dosing interval is considered effective for the beta-lactams. An effective T>MIC can be achieved by administering high doses at short intervals or with continuous infusions. Continuous infusions can maximize the time course of treatment and minimize fluctuations in serum concentrations. Continuous infusion regimens maintain trough levels and eliminate high peak serum concentrations which do not contribute to any additional benefit. Several studies report comparable outcomes with savings in the amount of drug administered.

Numerous animal studies indicate that varying drug concentrations and longer exposure time at or above MIC achieves maximum bacterial killing.

Macrolides, clindamycin, tetracyclines, and oxazolidinones achieve time-dependent bacterial killing with significant PAE. The AUC/MIC ratio predicts the efficacy. An AUC/MIC ratio of 25–35 for macrolides has been reported to achieve therapeutic efficacy against *Streptococcus pneumoniae*.

Concentration-Dependent Killing

C_{max}/MIC and AUC/MIC are predictors that correlate well with bactericidal efficacy.

Fluoroquinolones exhibit rapid concentration-dependent killing. The efficacy depends upon AUC/MIC or C_{max}/MIC ratio values. Few studies have documented the impact of PK/PD studies on clinical outcome in humans. Animal studies have demonstrated that AUC/MIC ratios of 25–30 for gram-positive organisms and 100–125 for gram-negative organisms are necessary to achieve successful clinical outcomes. Recent studies have suggested that the variability of pharmacodynamic parameters depends on specific pathogens.

Similar to the quinolones, aminoglycosides (e.g., gentamicin, tobramycin, and amikacin) exhibit rapid concentration-dependent killing. Experimentally, a high peak serum concentration of an aminoglycoside agent relative to MIC leads to faster killing of the bacteria. Achieving a C_{max}/MIC ratio of 8:1 against gram-negative organisms has been reported as a measure of success. Traditionally, PK monitoring was done to avoid serum levels associated with nephrotoxicity and dosage regimens were adjusted based on target serum concentrations. However, elevated C_{min} (trough) concentrations have been associated with toxicity. Administration of high doses for extended intervals (e.g., 5–7 mg/kg every 24 h for patients with normal renal function) is currently utilized by many practitioners to achieve enhanced bacterial killing and avoid the toxicity associated with higher C_{min}. With the extended dosing, serum concentrations fall below MIC for a period of time. All aminoglycosides demonstrate PAE which refer to suppression of bacterial growth despite zero serum concentration of the agent. PAE is more prevalent for aminoglycosides than any other class of anti-microbial agents against gram-negative bacilli. However, the role of PAE in achieving successful clinical outcomes is not established.

Pharmacokinetic Variability

Appropriate consideration of inter-patient variability is critical for designing an individual patient's anti-microbial therapy. Factors such as differences in absorption, distribution, metabolism, and elimination play a key role in the design. In addition, various disease states have a profound impact on PK parameters. For example, individualized weight based dosing for certain drugs such as beta-lactams, vancomycin, and fluoroquinolones would be appropriate for obese patients because of increased adipose tissue and volume results in an increased volume of distribution of the agent. Increased dosage of aminoglycosides is appropriate for burn patients because of increased renal perfusion and clearance.

Renal Impairment

The majority of the commonly used anti-microbial agents are renally excreted and it is critical to adjust the dosing in patients with renal impairment. The elderly population should be carefully monitored because of the approximate 5–10% reduction in glomerular filtration per decade beyond age of 30. Insufficient dosage adjustments play a key role in adverse events seen in the elderly. See Table 2.1 for a list of adverse events secondary to commonly used anti-microbials in the elderly.

Hepatic Impairment

Recommendations and methods for anti-microbial dosage adjustment in patients with renal dysfunction are well established. In contrast, this has not been established in drugs that are metabolized by the liver since the degree of impairment

Table 2.1 Adverse events of commonly used agents in the elderly

Class/agent	Adverse event
Aminoglycosides	Nephrotoxicity and ototoxicity
Beta-lactams	Fever, rash, thrombocytopenia, anemia, diarrhea, neutropenia
Carbapenems	Seizure
Clindamycin	Diarrhea, *Clostridium difficile*-associated colitis
Fluoroquinolones	CNS effects, decreased seizure threshold, nausea, vomiting
Linezolid	Thrombocytopenia, neutropenia
Tetracyclines	Photosensitivity
Trimethoprim–sulfamethoxazole	Drug fever, rash, blood dyscrasias

cannot be quantified by a single parameter. Adjustment in dosing schedules should be considered for agents significantly cleared by the hepatobiliary system. Patients with both hepatic and renal dysfunction are at greatest risk for potential adverse events since the elimination half-life of many agents is prolonged. Most importantly, PK trends are highly variable in this group rendering the evaluation of specific recommendations particularly difficult.

Critically Ill Patients

The PK parameters are altered and special consideration needs to be given for specific agent classes. For example, the inflammatory response associated with sepsis involves release of cytokines, endothelial damage, and changes in capillary permeability. These responses may in turn result in fluid shifts, third space losses, and decreased serum albumin levels. In addition, hepatic or renal impairment can result in prolonged half-life, decreased clearance, and drug accumulation.

For example, the volume of distribution of aminoglycosides can be significantly increased in critically ill (ICU) patients compared to non-critically ill patients with comparable renal function. As such, higher doses will be required to achieve therapeutic serum concentrations. For the same reason, aminoglycoside dosing nomograms cannot be used in this patient population. Adjustment of dosage will be required as patients recover from critical illness. Decreased serum albumin levels can result in adverse events for highly protein bound drugs, such as beta-lactams. Increased oral doses or use of intravenous antibiotics may be necessary for patients with decreased gastrointestinal perfusion and altered gastric emptying. Dosage regimens for beta-lactams or quinolones may require adjustment for patients with impaired renal perfusion and hepatic metabolism. On a similar note, a fluid load with an increase in drug volume of distribution may require a higher dosage regimen to address lower serum concentrations.

Continuous infusion of beta-lactam agents has been shown to produce more consistent serum levels than intermittent boluses. Oral fluoroquinolones have excellent bioavailability providing a great therapeutic management option in non-critically ill patients. However, administration in patients on enteral feeding may significantly impact absorption of the drug.

Tissue Concentrations

Many bacterial infections are extracellular. However, it is hard to determine the extent of antibiotic penetration in the interstitial area. Tissue penetration also depends upon the lipid solubility and plasma protein binding of the agent. For example, beta-lactams and aminoglycosides are hydrophilic and only the unbound free portion of the drug can penetrate outside the plasma. Protein binding can result in increased serum concentrations. Serum concentrations provide a better estimation of the tissue concentration for agents with low protein binding (aminoglycosides, fluoroquinolones) than those agents with higher tissue and protein binding. Localized infections, such as meningitis and osteomyelitis, pose a special challenge since many agents are unable to penetrate the area of infection.

Bacteriostatic Versus Bactericidal Drugs

The MIC or minimal inhibitory concentration is the level of drug that inhibits growth of an organism. This effect is traditionally measured by assessing turbidity of a culture. The MBC or minimal bactericidal concentration of a drug is the level at which the drug kills the bacterium (this is conventionally determined by plating a culture of bacteria on a plate without any anti-microbial agent at the MIC level when the culture is not turbid). As a general rule, cell wall active agents such as the beta-lactam antibiotics are bactericidal and agents that inhibit protein synthesis, like the tetracyclines or the macrolides are bacteriostatic and do not kill the organism in vitro. However, it is not always possible to determine whether an antibiotic will be bactericidal or bacteriostatic for a given organism. Agents that are predominantly bacteriostatic, like trimethoprim/sulfamethoxazole, may be bactericidal for certain organisms. A given agent may be bactericidal for one strain and bacteriostatic for another strain of the same bacterium. In most cases this has little relevance to patients, as bacteriostatic anti-microbials have an excellent track record of helping the host eliminate infections. There are some exceptions, the most prominent being the treatment of bacterial endocarditis. In this situation, clinical experience has indicated that bactericidal antibiotics are necessary to successfully treat the infection. This is thought to be related to the inability of polymorphonuclear leukocytes or other host cells to enter the fibrin clot that forms on the valve which is itself not vascularized. Other situations in which the use of a bactericidal antibiotic could be important are CNS infections and in the treatment of immunocompromised patients.

Conclusion

Our knowledge on PK/PD has evolved over the past two decades. Optimization of dose and duration of therapy via PK/PD tools provides patient specific therapy. It is essential to develop a process to successfully utilize PK/PD parameters which include collection of blood or sputum samples before initial therapy, isolation of pathogens, determination of MICs, and adjustment of initial therapy to match PK/PD parameters.

However, PK/PD studies are difficult to perform because of the challenges associated with obtaining culture samples, differentiating between individual patient defense mechanisms, time and cost. For the same reason, most of the PK/PD data are available from animal and in vitro studies. Another challenge is associated with our current process of obtaining unbound serum concentrations to monitor efficacy of an anti-microbial agent. Penetration of an agent into various sites of infection is variable and serum levels cannot accurately predict efficacy at the intracellular level. It makes sense to obtain individual PK and MIC data to optimize dosing.

Successful management of an infectious disease depends on various factors, including comorbidities, immune status, organ function, anti-bacterial activity, efficacy, tissue penetration, protein binding, and metabolism and elimination properties. Application of PK/PD data to establish optimal dosing regimens should lead to better patient outcomes and avoid escalation of anti-microbial resistance.

Further Reading

Eyler, R. F., & Mueller, B. A. (2010). Antibiotic pharmacokinetic and pharmacodynamic considerations in patients with kidney disease. Adv Chronic Kidney Dis, 17(5), 392-403.

Finberg, R. W., Moellering, R. C., Tally, F. P., Craig, W. A., Pankey, G. A., Dellinger, E. P., et al. (2004). The importance of bactericidal drugs: future directions in infectious disease. Clin Infect Dis, 39(9), 1314-1320.

Fry, D. E. (1996). The importance of antibiotic pharmacokinetics in critical illness. Am J Surg, 172(6A), 20S-25S.

Fry, D. E., & Pitcher, D. E. (1990). Antibiotic pharmacokinetics in surgery. Arch Surg, 125(11), 1490-1492.

Jacobs, M. R. (2001). Optimisation of antimicrobial therapy using pharmacokinetic and pharmacodynamic parameters. Clin Microbiol Infect, 7(11), 589-596.

Mehrotra, R., De Gaudio, R., & Palazzo, M. (2004). Antibiotic pharmacokinetic and pharmacodynamic considerations in critical illness. Intensive Care Med, 30(12), 2145-2156.

Roberts, J. A., & Lipman, J. (2009). Pharmacokinetic issues for antibiotics in the critically ill patient. Crit Care Med, 37(3), 840-851; quiz 859.

Rybak, M. J. (2006a). Pharmacodynamics: relation to antimicrobial resistance. Am J Infect Control, 34(5 Suppl 1), S38-45; discussion S64-73.

Rybak, M. J. (2006b). Pharmacodynamics: relation to antimicrobial resistance. Am J Med, 119 (6 Suppl 1), S37-44; discussion S62-70.

Thomas, J. K., Forrest, A., Bhavnani, S. M., Hyatt, J. M., Cheng, A., Ballow, C. H., et al. (1998). Pharmacodynamic evaluation of factors associated with the development of bacterial resistance in acutely ill patients during therapy. Antimicrob Agents Chemother, 42(3), 521-527.

Turnidge, J. D. (1998). The pharmacodynamics of beta-lactams. Clin Infect Dis, 27(1), 10-22.

Walsh, C. (2003). *Antibiotics: actions, origins, resistance*. Washington, D.C.: ASM Press.

Winter, M. E. (2004). *Basic clinical pharmacokinetics* (4th ed.). Philadelphia: Lippincott Williams & Wilkins.

Chapter 3
Sulfonamides

Introduction

Sulfonamides are anti-infective agents based on a sulfonamide group attached to a benzene ring (Fig. 3.1). Interestingly, the drugs were discovered by screening dyes for activity against streptococci using an animal model. The actual compound discovered by this method, Prontosil (the first anti-infective agent to be available commercially), has no in vitro activity and must be broken down in the animal or human host to sulfanilamide, the active agent.

Mechanism of Action

Sulfonamides are active against bacteria that cannot use external folate and must synthesize it from para-aminobenzoic acid (see Fig. 3.1). The sulfa drug is a para-aminobenzoic acid analog and serves as an antagonist of folate with relatively little toxicity to humans because humans can use folic acid without having to synthesize it from para-aminobenzoic acid. Thus the major toxicity for sulfa drugs relates to its potential as an allergen.

Although sulfonamides (of which many different compounds were synthesized) have a broad spectrum of activity against gram-positive organisms, and may be used to treat certain filamentous bacteria (e.g., *Nocardia*), their primary use today is in the treatment or prophylaxis of urinary tract infection caused by gram-negative organisms or in the treatment or prophylaxis to prevent *Pneumocystis jirovecii* or *Toxoplasma gondii* disease.

R.W. Finberg and R. Guharoy, *Clinical Use of Anti-infective Agents: A Guide on How to Prescribe Drugs Used to Treat Infections*,
DOI 10.1007/978-1-4614-1068-3_3,

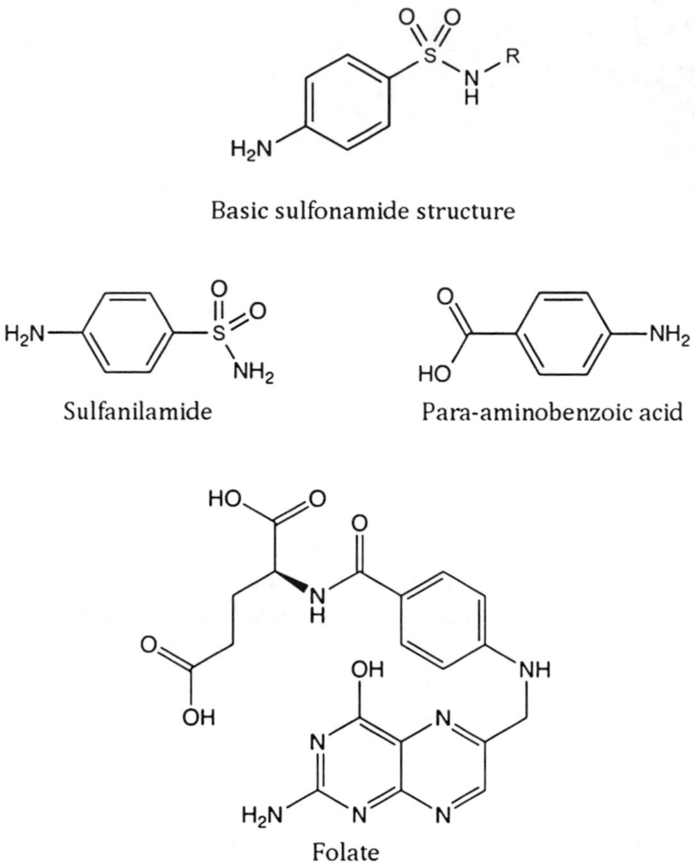

Basic sulfonamide structure

Sulfanilamide

Para-aminobenzoic acid

Folate

Fig. 3.1 Basic sulfonamide structure and structures of sulfanilamide, para-aminobenzoic acid, and folate

Trimethoprim

Trimethoprim is a trimethoxybenzylpyrimidine that inhibits bacterial dihydrofolic acid reductase. Dihydrofolate reductase converts dihydrofolic acid to tetrahydrofolic acid, which is a critical step in the synthesis of purines (Fig. 3.2). The lack of toxicity of this drug stems from its selectivity for the bacterial (as opposed to mammalian) dihydrofolate acid reductase. In this respect it is similar to the drug pyrimethamine which is used for the treatment of protozoan diseases (Fig. 3.3).

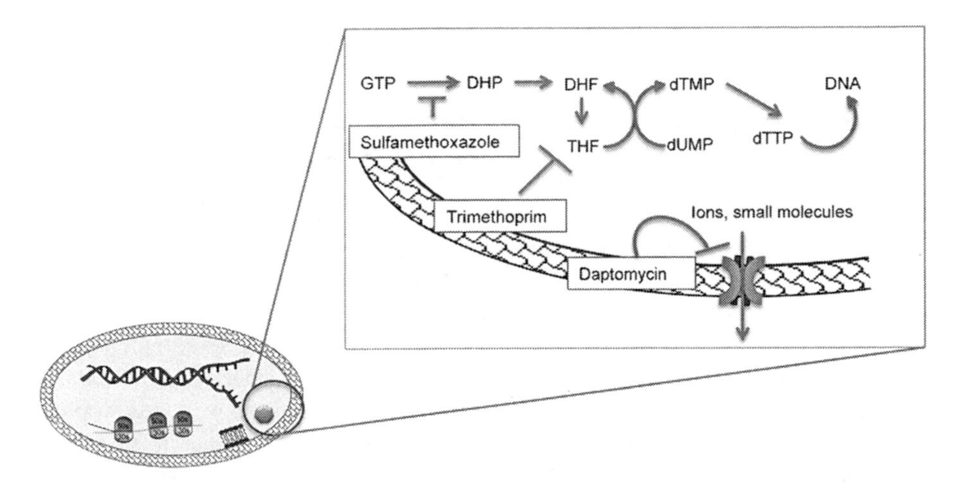

Fig. 3.2 Mechanism of action of agents that inhibit folate synthesis. *Abbreviations*: *GTP* guanosine triphosphate, *DHP* dihydropteroate, *DHF* 7,8-dihydrofolate, *THF* 5,6,7,8-tetrahydrofolate, *dUMP* 2′-deoxyuridine-5′-phosphate, *dTMP* 2′-deoxythymidine-5′-phosphate, thymidylate, *dTTP* deoxythymidine triphosphate, *DNA* deoxyribonucleic acid

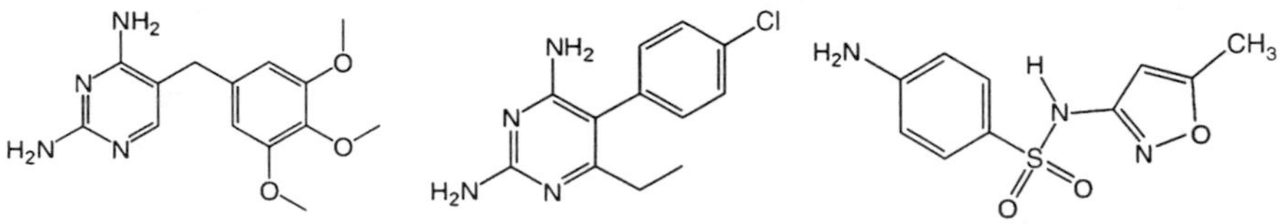

Trimethoprim Pyrimethamine Sulfamethoxazole

Fig. 3.3 Structures of trimethoprim, pyrimethamine, and sulfamethoxazole

Trimethoprim–Sulfamethoxazole Combinations

Trimethoprim–sulfamethoxazole (TMP/SMX) is a combination of a sulfa agent (sulfamethoxazole) that acts on the synthesis of folate, with a dihydrofolate reductase agent (trimethoprim) that acts on two different pathways of the synthesis of purines. These drugs demonstrate synergy as the combination of the two agents is far superior to the use of either one alone. In addition, while sulfa drugs are bacteriostatic agents (see Chap. 2), slowing growth but not killing the bacteria, the addition of trimethoprim leads to a combination drug that is frequently bactericidal, making this combination useful even when bactericidal activity is required. The fact that this drug is well absorbed and available both per oral (PO) and intravenous (IV) have made it useful for a variety of indications.

Spectrum of Activity

The combination of TMP/SMX has a broad spectrum of activity against aerobic gram-positive cocci and gram-negative rods. It can be used to treat common urinary pathogens including *E. coli*, *Klebsiella*, and *Enterobacter* species. It has no activity against *Pseudomonas* species. The combination is useful for treatment of sensitive *Shigella* and *Salmonella* species and also has activity against *Yersinia* and *Vibrio* species. In addition, the TMP/SMX combination has activity against *Streptococcus pneumoniae* and *Haemophilus influenzae*. It is well absorbed when given orally and has activity against sensitive strains of *S. aureus* including some methicillin-resistant *S. aureus* (MRSA) species. The agent has activity against *P. jirovecii* and *T. gondii* and is commonly used as both a first-line therapeutic as well as a prophylactic drug to prevent these diseases in immunocompromised patients. It is a useful agent in the treatment of immunocompromised patients because of its activity against *Listeria* and *Nocardia*.

Common Uses

In clinical practice today, most prescriptions are for the TMP/SMX combination. The combination is still commonly used for treatment of urinary tract infections and can be used for treatment of diarrhea illnesses caused by sensitive strains. A major use of the drug combination is in prophylaxis for *P. jirovecii* and *T. gondii* in immunocompromised hosts.

Common Uses of trimethoprim–sulfamethoxazole (Bactrim, Septra)
Treatment of *E. coli* urinary tract infections
Prophylaxis against urinary tract infections
Treatment of *P. jirovecii* and *T. gondii*
Prophylaxis against *P. jirovecii* and *T. gondii*

Adverse Effects

The most common toxicity of TMP/SMX, as noted above, is the development of an allergic response. It is estimated that 3% of the population may be allergic to sulfonamides. Although the manifestation of allergy is most commonly a simple skin rash, Stevens–Johnson syndrome (a severe defoliating rash with prominent mucous membrane involvement) and toxic epidermal necrolysis (TEN) are often associated with the use of sulfonamide drugs. Both of these syndromes can be fatal, and a history of either precludes use of sulfa drugs. Allergies are common in the general population, and even more common in patients with HIV-1 (where the drug is extensively used to prevent *P. jirovecii* and *T. gondii* disease). In addition, as the agent is a folate antagonist it can be associated with the development of anemia and leucopenia. TMP/SMX can be given with folinic acid (except in the treatment of enterococci, an organism that can use external folinic acid) to prevent these complications. In high doses TMP/SMX, may be associated with the development of hyperkalemia because of the effect of the trimethoprim component on renal tubular function. This is especially common in patients with renal impairment.

Further Reading

Huovinen, P., Sundstrom, L., Swedberg, G., & Skold, O. (1995). Trimethoprim and sulfonamide resistance. Antimicrob Agents Chemother, 39(2), 279–289.

Smilack, J. D. (1999). Trimethoprim-sulfamethoxazole. Mayo Clin Proc, 74(7), 730–734.

Chapter 4
Penicillins

Introduction

Historically, penicillins were the first "miracle drugs." When they first became commercially available, they were used to treat everything from severe staphylococcal wound infections to gonorrhea. Initially isolated as natural products of fungi, they have had an important role in medicine and have actually permanently altered the genetics of bacteria throughout the world. These drugs are characterized by the presence of the beta-lactam ring (Fig. 4.1), which is critical to their activity.

Mechanism of Action

Penicillins bind to penicillin-binding proteins (PBP), which are enzymes that are critical to the cross-linking of the cell wall of the bacterium. Beta-lactam antibiotics like penicillin bind to the active site of these enzymes that are critical in the transpeptidation reaction that results in peptidoglycan synthesis and cross-linking of the cell wall (Fig. 4.2). Without a rigid cell wall enclosing the bacterial cell membrane, the bacteria lose their shape and are susceptible to osmotic lysis. Binding of penicillins to PBP has been shown to trigger autolytic pathways (resulting in bacterial suicide) as well.

R.W. Finberg and R. Guharoy, *Clinical Use of Anti-infective Agents: A Guide on How to Prescribe Drugs Used to Treat Infections*,
DOI 10.1007/978-1-4614-1068-3_4, © Springer Science+Business Media, LLC 2012

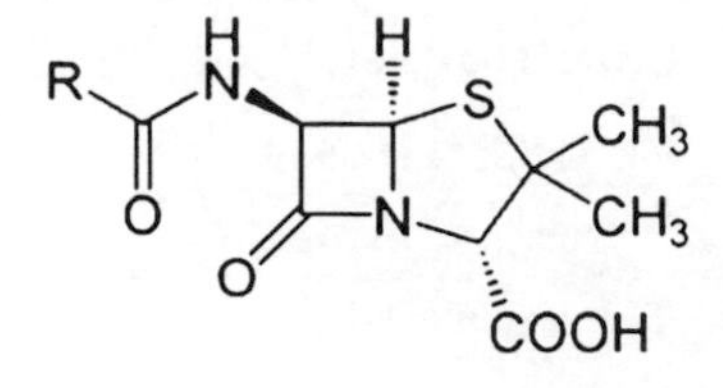

Fig. 4.1 Structure of penicillin

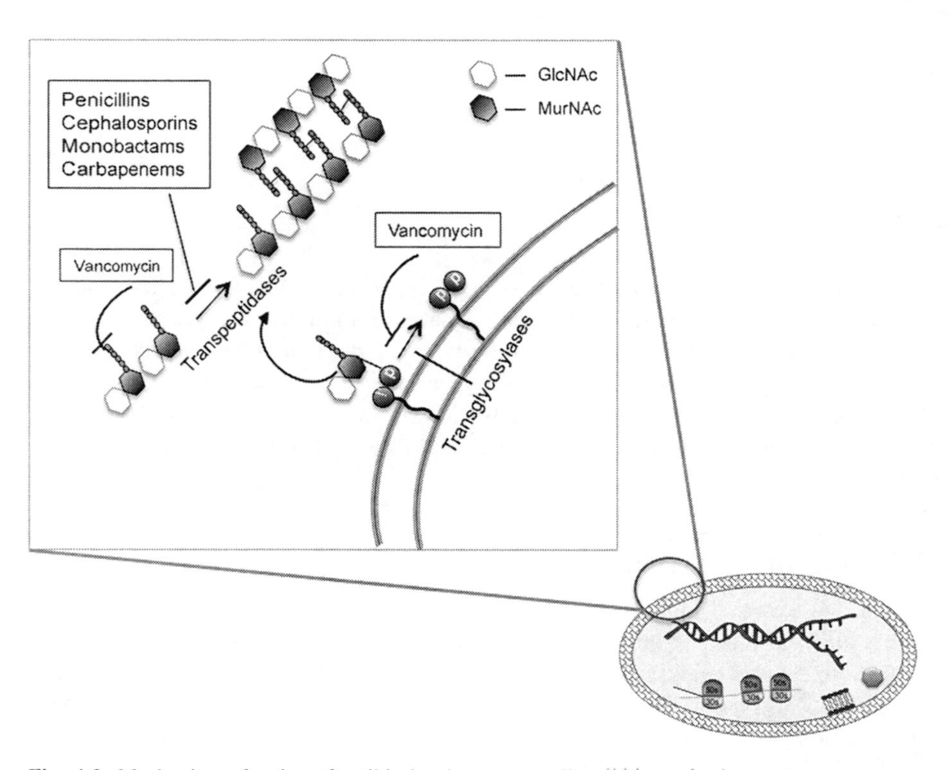

Fig. 4.2 Mechanism of action of antibiotics that act on cell wall biosynthesis

Classes of Penicillins

Natural Penicillins (Penicillin G and Penicillin V)

Penicillins G and V are the original penicillins (Fig. 4.3). They can be purified from cultures of the *Penicillium* mold. Penicillin G is acid-labile and therefore penicillin V is preferred for oral administration.

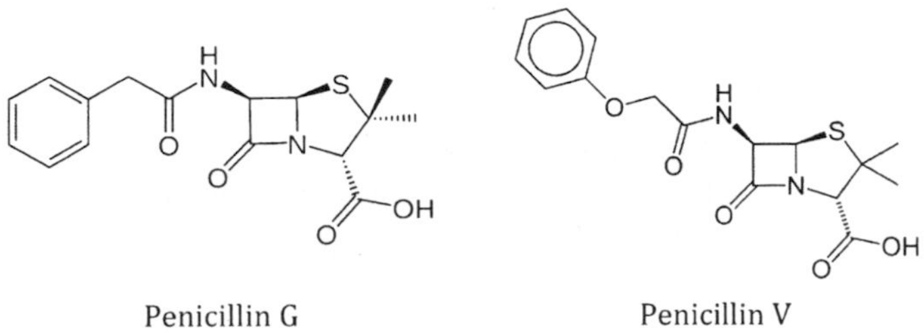

Penicillin G   Penicillin V

Fig. 4.3 Structure of Penicillin G and V

Spectrum of Activity

Since their activity is based on their ability to inhibit cell wall formation (and perhaps activate autolytic pathways), penicillins have a very broad spectrum of activity as all bacteria make cell walls. Certain organisms, such as syphilis and group A streptococci have never become resistant to penicillin. The gram-negative organism *Neisseria meningitidis* is also sensitive to the natural penicillins. Most anaerobic bacteria with the prominent exception of *Bacteroides fragilis* are sensitive to penicillin. *B. fragilis* produces a penicillinase that degrades the beta-lactam ring, making the drug ineffective against this organism. It is necessary to combine a penicillin with a beta-lactamase inhibitor (see below) to produce an anti-microbial with activity against *B. fragilis*. Although most *Clostridium* species (e.g., *C. perfrigens* and *C. diphtheria*) are sensitive to penicillin G, *C. difficile* (a common cause of antibiotic-associated diarrhea and a possible cause of severe colitis) is completely resistant to all penicillins.

While most strains of *Staphylococcus aureus* were sensitive to penicillin G when it was first introduced into clinical practice in the 1940s, these bacteria rapidly acquired the ability to produce beta-lactamases. While the transition to *S. aureus* strains that were resistant to penicillin G began in hospitals where penicillins were first used, it has spread throughout the world and most strains of *S. aureus* now produce staphylococcal beta-lactamases and therefore are now treated with anti-staphylococcal penicillins (see below).

Common Uses

Common uses of penicillin G include treatment of syphilis, group A streptococcal infections, *Leptospira* and *Borrelia* infections, and clostridial infections (except *C. difficile*). Most strains of *S. pneumoniae* (the cause of most cases of bacterial pneumonia and bacterial meningitis) are sensitive to penicillin G (Table 4.1). However, strains of *S. pneumoniae* with various levels of resistance to penicillin have been

Table 4.1 Common uses for penicillins

Penicillin	Common uses
Pen G	Syphilis, leptospirosis, *Borrelia* infections, yaws
Pen G	Anaerobic lung abscess
Pen G	Streptococcal infections (including viridans streptococcal endocarditis and all group A streptococcus infections)

reported, making a third-generation cephalosporin or vancomycin (see below), the most common initial choices for treatment of these common diseases. While the growth of enterococci is inhibited by the use of these agents, penicillins alone are not bactericidal for enterococci and a combination of penicillin and an aminoglycoside is required for treatment of enterococcal endocarditis. An aminopenicillin (see below) can be used to treat urinary tract infections caused by enterococci) (Table 4.1).

Anti-staphylococcal Penicillins: Methicillin, Oxacillin, and Nafcillin

The anti-staphylococcal penicillins differ from the natural penicillins by the addition of a side chain that prevents binding by the staphylococcal beta-lactamases.

The presence of a bulky side chain makes these drugs excellent choices for treatment of many *S. aureus* strains that are penicillin-resistant. While methicillin is the drug most commonly used in sensitivity testing, it is no longer marketed because of a high incidence of interstitial nephritis (Fig. 4.4).

Nafcillin, which can cause phlebitis, and oxacillin, which has been associated with hepatitis, are used instead of methicillin (Fig. 4.5). Nafcillin and oxacillin are used preferentially for *S. aureus* (in place of penicillin G) but neither has activity against MRSA, as these organisms have alterations in their PBP which render them resistant to all penicillins (treatment with vancomycin or other agents is required – see below). Streptococci including pneumococci are less sensitive to the anti-staphylococcal penicillins than they are to penicillin G and V.

Common uses for anti-staphylococcal penicillins (nafcillin and oxacillin) include skin infections and deep-seated abscesses and wound infections caused by *S. aureus*. As these drugs are bactericidal for most strains of *S. aureus* (other than MRSA), they can be used to treat endocarditis. Cerebrospinal fluid (CSF) penetration, as with most penicillins, is poor but they do cross the blood–brain barrier when the meninges are inflamed. Therefore these drugs will attain levels that are adequate for treatment of *S. aureus* meningitis in the case of infections caused by sensitive strains.

Cloxacillin and dicloxacillin are anti-staphylococcal agents formulated for oral administration.

Fig. 4.4 Structure of methicillin

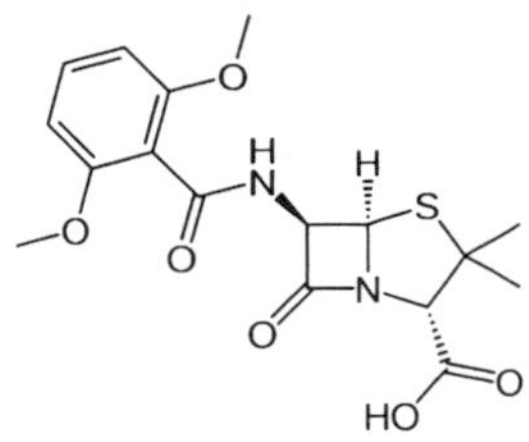

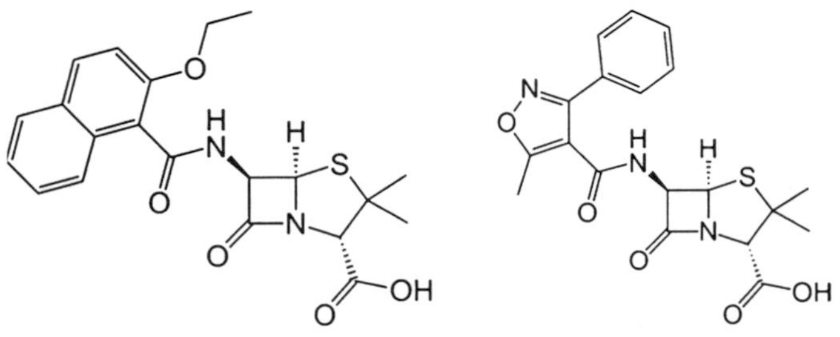

Nafcillin

Oxacillin

Fig. 4.5 Structure of anti-staphylococcal penicillins, nafcillin and oxacillin

Common Uses for Anti-staphylococcal Penicillins

Anti-staphylococcal penicillins are commonly used for the treatment of *S. aureus* skin, blood, and respiratory infections (except MRSA).

Aminopenicillins

One of the defenses of gram-negative bacteria against the natural (and anti-staphylococcal) penicillins is the presence of porins in the outer membrane of the bacteria that prevent the hydrophobic natural penicillins from penetrating into the periplasmic space. The addition of an amino group allows the drug to penetrate into the periplasmic space of certain gram-negative organisms. This gives ampicillin and amoxicillin activity against *Proteus mirabilis*, *H. influenzae*, and *E. coli*, as well as *Salmonella* and *Shigella*. As noted above, ampicillin and amoxicillin, unlike penicillin, are often used for treatment of enterococcal infections, especially urinary tract infections. These agents are also commonly used for treatment of *Listeria* infections as well as pneumonia and meningitis caused by susceptible strains of *H. influenzae* (Fig. 4.6).

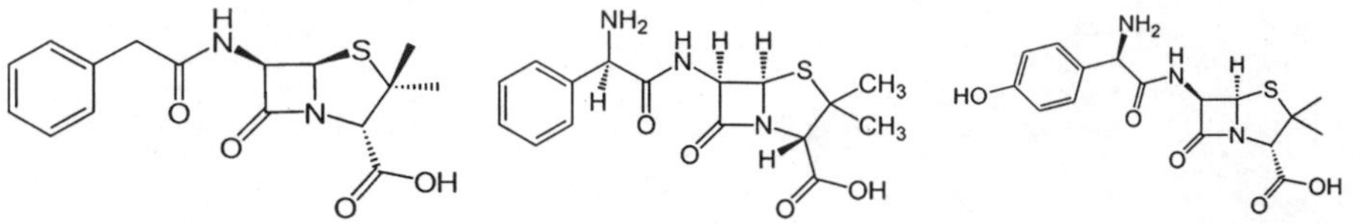

Benzyl penicillin Ampicillin Amoxicillin

Fig. 4.6 Structure of benzyl penicillin, ampicillin, and amoxicillin

Spectrum of Activity of Aminopenicillins

The aminopenicillins retain the gram-positive and anaerobic spectrum of penicillin and can be used to treat anaerobic infections of the respiratory system. Like penicillin, they are sensitive to beta-lactamases produced by *B. fragilis* and cannot be used for anaerobic infection in the lower GI tract. In addition to the spectrum of penicillins, which includes all group A streptococci and most streptococci, ampicillin and amoxicillin have an additional spectrum that includes *E. coli*, *P. mirabilis*, and *Enterococcus* species so that they are commonly used to treat urinary tract infections. The aminopenicillins also have activity against sensitive strains of *Shigella* and *Salmonella*, although treatment of *Salmonella* may prolong the carrier state in some cases. They are useful for treatment of *Listeria*.

Common Uses of Aminopenicillins

Treatment of urinary tract infections (*E. coli* and *Enterococcus*)
Treatment of otitis media
Treatment of serious enterococcal infections
Treatment of *Listeria* meningitis
Treatment of endocarditis

While ampicillin is available as both a parenteral (usually given IV) and oral form, amoxicillin is usually preferred as an oral agent because of its better absorption and longer half-life (requiring three times a day dosing as opposed to four times a day dosing).

Combinations of Aminopenicillins with Beta-Lactamase Inhibitors

While the anti-staphylococcal penicillins are effective in treating gram-positive organisms that produce beta-lactamases, they are ineffective against beta-lactamases produced by gram-negative organisms. To extend the spectrum of the amino penicillins, they are often paired with a beta-lactamase inhibitor. The addition of

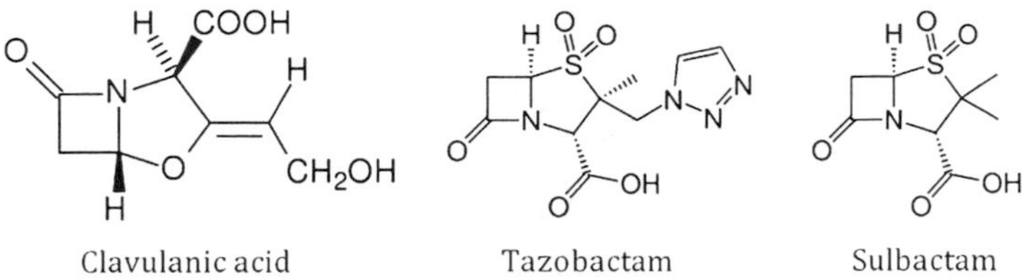

Clavulanic acid Tazobactam Sulbactam

Fig. 4.7 Structure of beta-lactamases: clavulanic acid, tazobactam, and sulbactam

sulbactam to ampicillin (ampicillin/sulbactam or Unasyn) extends the spectrum of ampicillin to include *B. fragilis*, as well as methicillin-sensitive staph, and a broad spectrum of gram-negative aerobes (including most *Enterobacteriaceae*, such as *Enterobacter*, *Klebsiella*, *Proteus species*, and *Serratia*). For PO administration, amoxicillin is paired with clavulanic acid (amoxicillin/clavulanate or Augmentin). Both of these drugs are useful agents for treating infections that may involve both aerobes and anaerobes (Fig. 4.7).

Spectrum of Activity of Aminopenicillin/Beta-Lactamase Combinations

In addition to the spectrum of the aminopenicillins, the addition of clavulanic acid or sulbactam adds activity against organisms that produce beta-lactamases, including *S. aureus* and *B. fragilis* and some gram-negatives which are resistant to ampicillin on the basis of their production of beta-lactamase. They retain the activity of ampicillin against *H. influenzae*, *P. mirabilis*, and most anaerobes but add the advantage of activity against *B. fragilis* and some resistant pneumococcus and *H. influenzae* making them better drugs for the treatment of otitis media in this era of resistant *H. influenzae* and *S. pneumoniae*. These drugs have no activity against *Pseudomonas aeruginosa*.

Common Uses for Aminopenicillin/Beta-Lactamase Combinations

Otitis media
Cholangitis
Septic arthritis
Animal or human bite wounds

Extended-Spectrum Penicillins

Pseudomonas species, in addition to having porins that restrict the entry of penicillins into the periplasmic spaces and having beta-lactamases in the periplasmic space, express "pumps" that can pump the antibiotics into the periplasmic spaces so that

Fig. 4.8 Structure of piperacillin

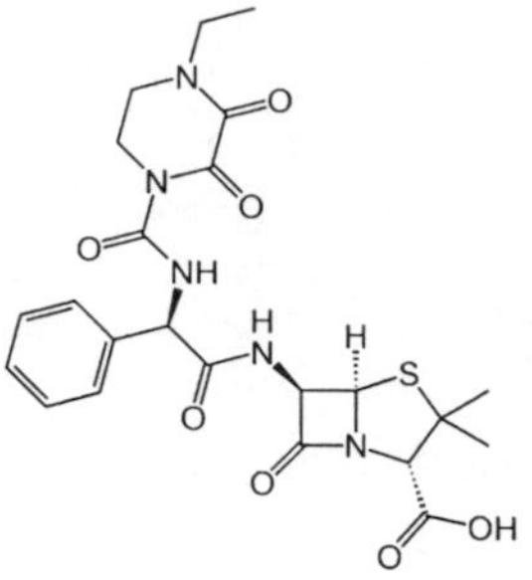

they do not affect cell wall synthesis. The extended-spectrum penicillins have side chains that allow them to better penetrate the gram-negative porins and thus accumulate at higher levels in the periplasmic space. These agents (carbenicillin, ticarcillin, and piperacillin) are also relatively more resistant to beta-lactamases than the aminopenicillins. This combination of better penetration into the periplasmic spaces and resistance to beta-lactamases gives these agents activity against *Pseudomonas* species in addition to activity against most *Enterobacteriaciae* (Fig. 4.8).

Spectrum of Activity of Extended-Spectrum Penicillins

These agents are useful for the treatment of a broad variety of gram-positive cocci (including strep and non-beta-lactamase producing staph) and gram-negative rods (including most aerobic gram-negative organisms and most anaerobes other than *B. fragilis*).

Common Uses of Extended-Spectrum Penicillins

Extended-Spectrum penicillins are used for pseudomonal infections including urinary tract infections, skin infections, and pneumonia.

Extended-Spectrum Penicillins Combined with Beta-Lactamase Inhibitors

Despite the broad spectrum of the extended-spectrum penicillins, they are still ineffective against beta-lactamase-producing *S. aureus* and beta-lactamase-producing anaerobic organisms (*B. fragilis*). To extend the spectrum to include these organisms they are paired with beta-lactamase inhibitors to produce two drug combinations: piperacillin/tazobactam (Zosyn) and ticarcillin/clavulanate (Timentin). These parenteral agents have a spectrum including almost all anaerobes (except *C. difficile*) as well as most aerobic gram-positive organisms including *S. aureus* (but not MRSA) and most aerobic gram-negative organisms.

Spectrum of Activity of Extended-Spectrum Penicillins Combined with Beta-Lactamases

The extended-spectrum penicillin/beta-lactamase combination drugs have among the broadest spectrum of activity of any available agents. They can be used to treat *Pseudomonas aeruginosa* infections in addition to most other gram-negative aerobic infections. They have excellent activity against gram-positive cocci and rods, including beta-lactamase-producing *S. aureus*.

Common Uses of Extended-Spectrum Penicillin Combined with Beta-Lactamases

Extended-spectrum penicillin in combination with beta-lactamases has been used for ventilator or in-hospital acquired pneumonia as well as abdominal sepsis in situations where the infecting organism is not known or there are multiple organisms causing diseases.

Adverse Effects and Toxicities of Penicillins

While side effects such as nausea and vomiting can occur with administration of penicillins, the most worrisome side effects relate to the development of anaphylactic reactions. Although allergies to penicillins are reported to occur in up to 10% of the population, the incidence of anaphylactic reactions is considerably lower (0.004–0.015%). It is noteworthy that allergic responses are over-reported, and while cross-reactivity between penicillin and first cephalosporins probably does occur (see Chap. 5), cross-reactivity between penicillins and third-generation cephalosporins probably is uncommon (see Chap. 15).

Other side effects that can occur from administration of penicillins include rashes, drug fever, urticaria, interstitial nephritis, liver function test abnormalities (particularly common with oxacillin), bone marrow suppression (especially neutropenia), and serum sickness. Administration of penicillins at very high doses can cause abnormalities of platelet function, and administration of penicillins in high doses in the presence of renal failure has been associated with CNS toxicity.

Further Reading

Drawz, S. M., & Bonomo, R. A. (2010). Three decades of beta-lactamase inhibitors. Clin Microbiol Rev, 23(1), 160–201.

Pichichero, M. E. (2005). A review of evidence supporting the American Academy of Pediatrics recommendation for prescribing cephalosporin antibiotics for penicillin-allergic patients. Pediatrics, 115(4), 1048–1057.

Chapter 5
Cephalosporins

Introduction

After the success of the penicillins, the next logical development in antibiotic therapy was the discovery of the cephalosporins. These agents were also isolated from fungi (originally from the fungus *Cephalosporium acremonium*) and were noted, early in their history, to be effective against organisms that produced beta-lactamases.

Mechanism of Action

Like penicillins, the cephalosporins attach to penicillin-binding proteins and inhibit the synthesis of peptidoglycan and therefore impair cell wall synthesis (see Fig. 4.2). Like penicillins, they are concentration-dependent antibiotics and best results are obtained if they have levels above the MIC (minimal inhibitory concentration) of the organism at all times. Thus administration by continuous infusion would be logical.

By virtue of their resistance to beta-lactamases, cephalosporins have activity against most *S. aureus* strains and against some gram-negative organisms (particularly *Escherichia coli*, *Klebsiella pneumoniae*, and *Proteus mirabilis*). Cephalosporins differ from penicillins in having a six-membered dihydrothiazine ring. The six-membered ring of the cephalosporin family (Fig. 5.1) is much less stable than the five-membered ring of penicillins. This has important implications in terms of the likelihood of allergies to cephalosporins in penicillin allergic patients. While degradation of the penicillin ring leads to a stable penicilloate ring, the cephalosporins' rings are rapidly broken down into fragments. These differences in processing suggest that cross-reactive allergic responses are likely to be minimal unless the side chains are similar (see Chap. 15).

R.W. Finberg and R. Guharoy, *Clinical Use of Anti-infective Agents: A Guide on How to Prescribe Drugs Used to Treat Infections*,
DOI 10.1007/978-1-4614-1068-3_5,

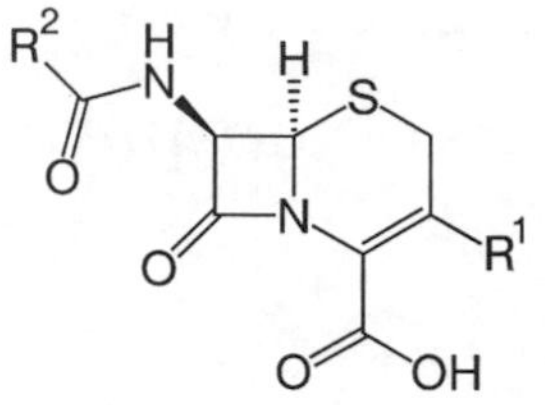

Fig. 5.1 Structure of cephalosporins

Classes of Cephalosporins

The different cephalosporins are classified by "generation" depending on their date of introduction into the market. In general, the progressive generations are marked by increased activity against a broader and broader range of gram-negative organisms. The third generation cephalosporins are notable for their ability to cross the blood–brain barrier while first and second generation cephalosporins do not cross well and consequently cannot be used to treat patients with meningitis (Table 5.1).

A couple of words of caution are given with regard to cephalosporins (1) unlike the penicillins, many of which have activity against *Enterococcus*, cephalosporins have no activity against enterococci, (2) with the exception of the second generation cephalosporins (cefoxitin and cefotetan), these agents do not have activity against *B. fragilis*.

1. *First generation cephalosporin*: The first cephalosporin to be extensively marketed, cephalothin, was used extensively for *S. aureus* infections as well as for *E. coli* and *K. pneumoniae* infections. Cephalothin, which is not highly protein bound, but has a short half-life, has been abandoned in the US in favor of cefazolin, an antibiotic with a similar spectrum of activity but with a longer half-life. Cefazolin (Fig. 5.2), unlike cephalothin, can be given intra-muscularly in addition to intravenously.

   The first generation cephalosporins were marketed, in part, for their ability to treat *S. aureus* which had become resistant to penicillin. In addition to gram-positive organisms, however, the spectrum of cephalothin includes several gram-negative organisms, especially *E. coli* and *Klebsiella.* The spectrum includes most organisms killed by penicillin (including *Streptococcus pyogenes*, viridans streptococci, and *S. pneumoniae* but the cephalosporins are not able to kill *Enterococcus*).

   Cefazolin, because of its gram-positive and gram-negative spectra, is one of the most commonly used antibiotics. Cefazolin is often used to treat skin infections because of its activity against both strep and staph. It is often used in place of penicillin in the case of patients with a history of penicillin allergy, although these patients may have allergic responses to cefazolin (a first generation cephalosporin) as well because of the similarity in the side chains of penicillins and first generation cephalosporins (see Chap. 15). Because of its relatively long half-life and ease of administration, cefazolin is commonly used as a prophylactic agent to prevent surgical infections. In this situation the antibiotic is administered

Table 5.1 Classes and uses of cephalosporins

Generation	Example agents	Spectrum and Use
First	Cefazolin	Staph, strep, *E. coli*, commonly used for surgical prophylaxis
Second	Cefotetan	Adds *B. fragilis* activity
Third	Ceftriaxone, Cefotaxime, Ceftazidime	Broad spectrum of gram-negatives, ceftriaxone and cefotaxime used for meningitis, ceftazidime has *P. aeruginosa* activity
Fourth	Cefepime	Better staph activity than ceftazidime
Fifth	Ceftaroline	Activity against MRSA

Fig. 5.2 Structure of cefazolin

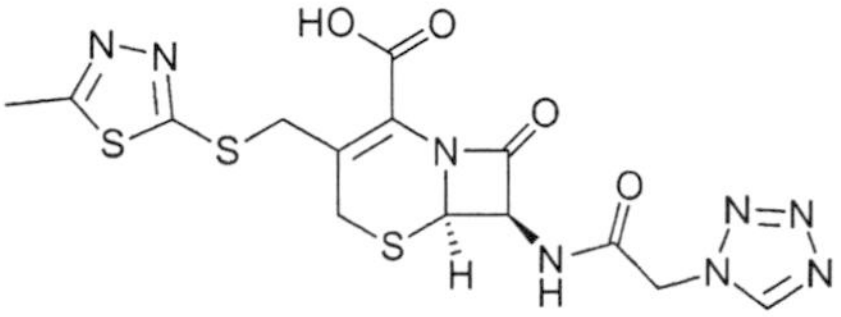

Fig. 5.3 Structure of cefotetan

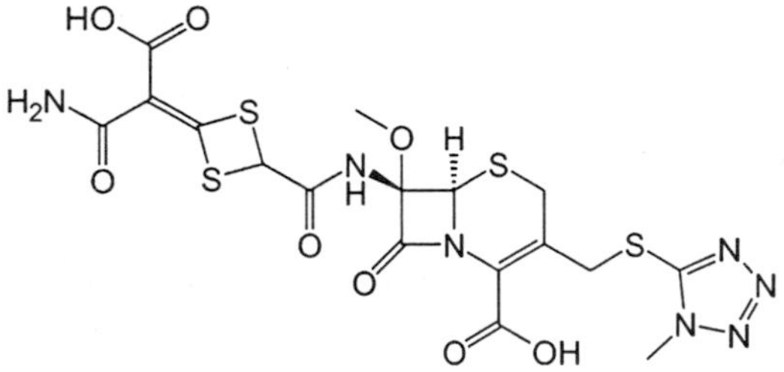

prior to surgery. Experimental animal studies indicate that the administration of an antibiotic prior to opening a wound decreases the risk of wound infection, but the major effect of cefazolin in the case of "clean surgery" is to prevent urinary tract infections, in particular *E. coli* infections.

2. *Second generation cephalosporins*: The second generation cephalosporins include the true cephalosporins, cefamandole, and cefuroxime, as well as the cephamycins, cefoxitin, and cefotetan. The cephamycins, while chemically and pharmacologically similar to the cephalosporins (accounting for their being grouped together), are derived from the bacterium *Streptomyces,* as opposed to the true cephalosporins which are derived from the fungus *Cephalosporium acremonium.* Cefoxitin and cefotetan have an additional methoxy group in the beta-lactam ring (Fig. 5.3).

   This makes these drugs less sensitive to the beta-lactamase produced by anaerobes like *B. fragilis.* For this reason they can be used to treat abdominal infections including situations in which *B. fragilis* is a major component of the infecting flora. Despite the fact that these agents have a broader gram-negative spectrum, including better activity against most *E. coli, Klebsiella*, and *Proteus* species, their activity against staphylococci and streptococci is decreased.

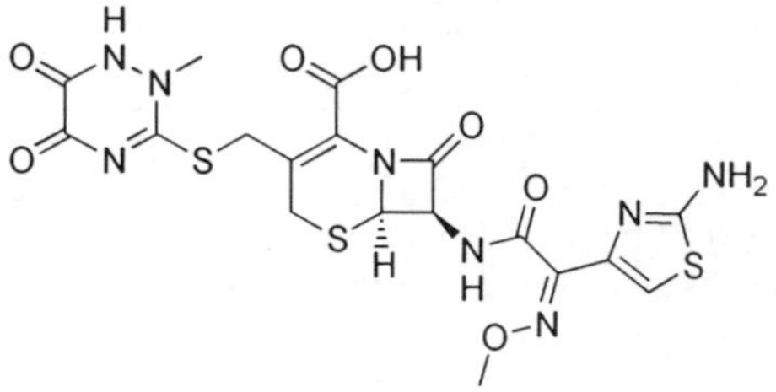

Fig. 5.4 Structure of ceftriaxone

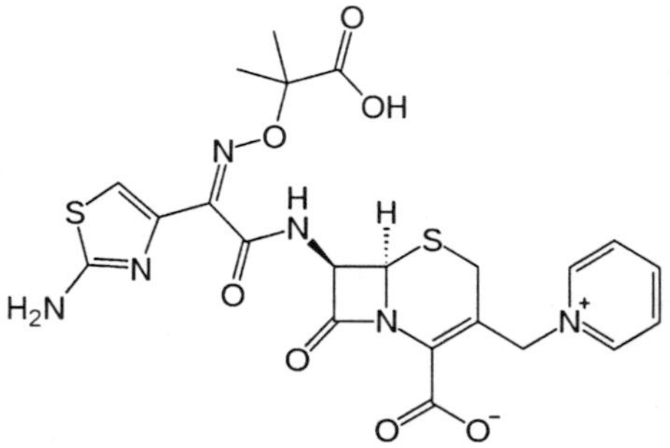

Fig. 5.5 Structure of ceftazidime

The true second generation cephalosporins (cefamandole and cefuroxime) have activity not only against a broader spectrum of *Enterobacteriaceae* than do the first generation cephalosporins, but they also have activity against *H. influenzae*. Cefamandole is no longer available in the US and although cefuroxime can be used to treat *H. influenzae* infections, the meningeal penetration of the third generation cephalosporins (ceftriaxone or cefotaxime), make these agents preferable if meningeal disease is considered.

3. *Third generation cephalosporins*: The third generation cephalosporins represent a major advance in the treatment of meningitis. Unlike previous generations of cephalosporins, these agents have the ability to cross the blood–brain barrier. This results in their routine use as a first-line agent against bacterial meningitis since they have activity against *S. pneumoniae* and *H. influenzae* as well as *N. meningitidis*. Ceftriaxone, because of its long half-life (5.8–8.7 h) can be given as a single daily dose for many indications (the exception being meningitis where it is conventionally given at maximum doses: 2 g IV q12H for adults (Fig. 5.4).

   Cefotaxime also has good CSF penetration and activity against organisms likely to cause meningitis but must be given every 4 h for optimal activity in meningitis).

   While ceftriaxone, cefotaxime, and ceftizoxime all have activity against a broad range of both gram-negative and gram-positive organisms, they are not optimal drugs for treating *Pseudomonas aeruginosa*. A modification of the R1 side chain to produce ceftazidime (Fig. 5.5) results in better activity against

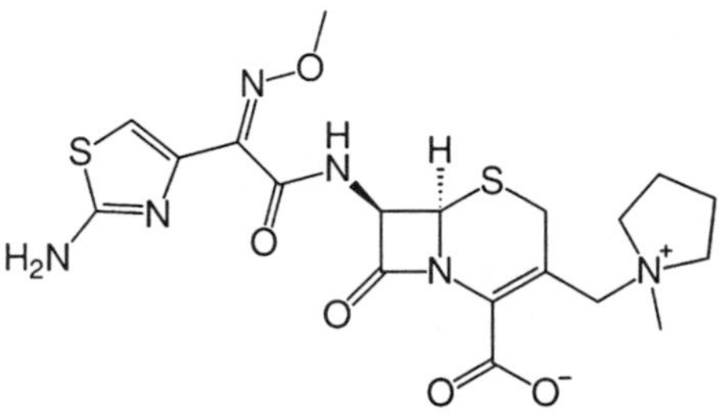

Fig. 5.6 Structure of cefepime

P. aeruginosa. Unfortunately, this modification results in less affinity for the penicillin-binding proteins of staphylococci, resulting in less activity against staphylococci. Because of its broad range of activity, ceftazidime has been used extensively for treatment of infections in patients lacking polymorphonuclear leukocytes (neutropenic patients – usually people with acute leukemia actively receiving chemotherapy).

4. *Fourth generation cephalosporins*: A synthetic modification of the R2 side chain of third generation cephalosporins results in the synthesis of cefepime (Fig. 5.6). This drug maintains the staph activity that was lost with ceftazidime. In addition, cefepime has activity against some chromosomally encoded beta-lactamases, giving a broader gram-negative and gram-positive spectrum than ceftazidime. Like ceftazidime it has little anaerobic activity.
5. *Ceftaroline*: Ceftaroline is a new cephalosporin with activity against MRSA and resistant *S. pneumoniae*. It has demonstrated efficacy in the treatment of complicated skin and soft tissues infections where MRSA is a likely pathogen and in community-acquired pneumonia where resistant *S. pneumoniae* is a possible pathogen. Ceftaroline 600 mg IV q12H has been demonstrated to have clinical efficacy equivalent to the combination of vancomycin and aztreonam for the treatment of skin and soft tissue infections and is equivalent to ceftriaxone in the treatment of community-acquired pneumonia. The drug is renally excreted so dosage adjustment is required for patients with renal failure.

Adverse Effects

As with other beta-lactam antibiotics, toxic effects are unusual. The most common causes of cephalosporin toxicity relate to their ability to cause allergic reactions including skin rashes and idiosyncratic reactions. Abnormalities of liver function tests have been reported with treatment. As with most other drugs, they may cause reversible bone marrow suppression.

Common Uses of Cephalosporins (Table 5.2)

Table 5.2 Common uses of cephalosporins

Cephalosporin	Common Uses
Cefazolin	Prophylaxis – given before most surgical procedures
Ceftriaxone	Commonly used for meningitis
	Endocarditis
	Treatment of gonorrhea
Ceftazidime	Treatment of neutropenic patients with fever
Ceftaroline	Treatment of MRSA skin infections

Further Reading

Dancer, S. J. (2001). The problem with cephalosporins. J Antimicrob Chemother, 48(4), 463–478.

Kalman, D., & Barriere, S. L. (1990). Review of the pharmacology, pharmacokinetics, and clinical use of cephalosporins. Tex Heart Inst J, 17(3), 203–215.

Prasad, K., Kumar, A., Gupta, P. K., & Singhal, T. (2007). Third generation cephalosporins versus conventional antibiotics for treating acute bacterial meningitis. Cochrane Database Syst Rev(4), CD001832.

Steed, M. E., & Rybak, M. J. (2010). Ceftaroline: a new cephalosporin with activity against resistant gram-positive pathogens. Pharmacotherapy, 30(4), 375–389.

Chapter 6
Monobactams

Aztreonam

Introduction

Aztreonam is the only synthetic monobactam available in the US. It is isolated from *Chromobacterium violaceum*. In contrast to bicyclic beta-lactams such as penicillins and cephalosporins, it has a sulfonic group at the N-1 position (Fig. 6.1).

Mechanism of Action

The sulfonic acid group at the N-1 position activates the beta-lactam ring assisting acetylation of transpeptidases. The result is inhibited synthesis of bacterial cell walls (see Fig. 4.2). Aztreonam binds to penicillin-binding protein 3 of susceptible gram-negative pathogens forming elongated filamentous cells which ultimately lyse, resulting in death. It does not bind to the penicillin-binding proteins of gram-positive organisms, rendering it ineffective for treatment of gram-positive infections. In addition, it is not effective against anaerobic pathogens.

Spectrum of Activity

Aztreonam activity is limited to aerobic gram-negative bacilli since it lacks affinity for penicillin-binding proteins of gram-positive bacteria and anaerobic organisms. The iminopropyl carboxyl group on the side chain contributes to beta-lactamase stability against *Pseudomonas aeruginosa* and the aminothiazolyl side chain contributes to activity against gram-negative bacteria.

R.W. Finberg and R. Guharoy, *Clinical Use of Anti-infective Agents: A Guide on How to Prescribe Drugs Used to Treat Infections*,
DOI 10.1007/978-1-4614-1068-3_6, © Springer Science+Business Media, LLC 2012

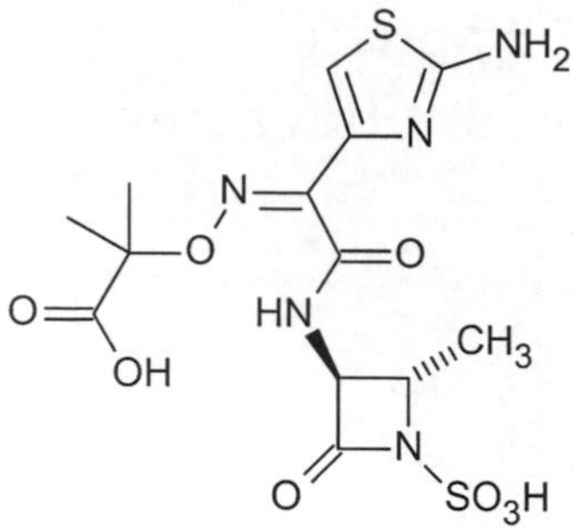

Fig. 6.1 Structure of aztreonam

Aztreonam is active against *N. gonorrhoeae*, *N. meningitidis*, *M. catarrhalis*, *H. influenzae*, *E. coli*, *Klebsiella* spp. (except ESBL-producing), *Enterobacter*, *Serratia*, *Shigella*, *Proteus mirabilis*, *Proteus vulgaris*, *Providencia*, *Morganella*, *Citrobacter*, *Aeromonas*, *P. aeruginosa* and *Pasteurella multocida*.

Common Uses

Because of its narrow spectrum, aztreonam is usually combined with another agent to treat serious infections. It is used for treatment of infections of the skin, soft tissue, bone and joint, treatment of lower respiratory tract infections (pneumonia), and for pelvic, intra-abdominal, and urogenital infections. It can also be used via inhalation to treat *P. aeruginosa* infections in patients with cystic fibrosis. The kidney excretes 60–70% of aztreonam and hence dosage adjustment is advised for patients with impaired renal function.

Adverse Effects

Common side effects include reaction(s) at the site of injection, rash, diarrhea, nausea, and vomiting. In a small number of patients nephrotoxicity (higher incidents when receiving concomitant nephrotoxic agents), neutropenia, thrombocytopenia, electrocardiogram abnormality, reaction at injection site, rash, urticaria, headache, confusion and seizure have been reported. Abdominal pain, bronchospasm, wheezing and chest discomfort have been reported with inhalation use.

Further Reading

Brewer, N. S., & Hellinger, W. C. (1991). The monobactams. Mayo Clin Proc, 66(11), 1152–1157.

Clark, P. (1992). Aztreonam. Obstet Gynecol Clin North Am, 19(3), 519–528.

Cunha, B. A. (1993). Aztreonam. Urology, 41(3), 249–258.

Ennis, D. M., & Cobbs, C. G. (1995). The newer cephalosporins. Aztreonam and imipenem. Infect Dis Clin North Am, 9(3), 687–713.

Johnson, D. H., & Cunha, B. A. (1995). Aztreonam. Med Clin North Am, 79(4), 733–743.

Chapter 7
Carbapenems

Introduction

The carbapenems are beta-lactam agents that are distinguished from penicillin by the substitution of a carbon atom for a sulfur atom and the addition of a double bond to the five-membered ring of the penicillin nucleus. Imipenem, meropenem, ertapenem, and doripenem are commercially available in the United States (Fig. 7.1). They are derived from *Streptomyces catteleya* and offer a broader spectrum of activity than most other beta-lactam agents. The broad-spectrum activity of carbapenems is believed to be due to three factors: (1) affinity for penicillin-binding proteins (peptidases) from a wide range of bacteria; (2) ability to penetrate the cell membrane of multiple gram-negative bacilli; and (3) resistance to a broad range of beta-lactamases from gram-positive and gram-negative bacilli.

The carbapenems are eliminated by glomerular filtration requiring dosage adjustment in renally impaired or elderly patients.

Mechanism of Action

Like other beta-lactams, the carbapenems bind to penicillin-binding proteins, disrupting bacterial cell wall synthesis (see Fig. 4.2) and killing susceptible microorganisms. They are resistant to hydrolysis by most beta-lactamases.

Spectrum of Activity

Imipenem, meropenem, and doripenem are highly active against a broad array of aerobic and anaerobic organisms including staphylococci (excluding MRSA), *Listeria* spp., enterococci (excluding *E. faecium*), streptococci (including *S. pneumoniae*),

R.W. Finberg and R. Guharoy, *Clinical Use of Anti-infective Agents: A Guide on How to Prescribe Drugs Used to Treat Infections*,
DOI 10.1007/978-1-4614-1068-3_7,

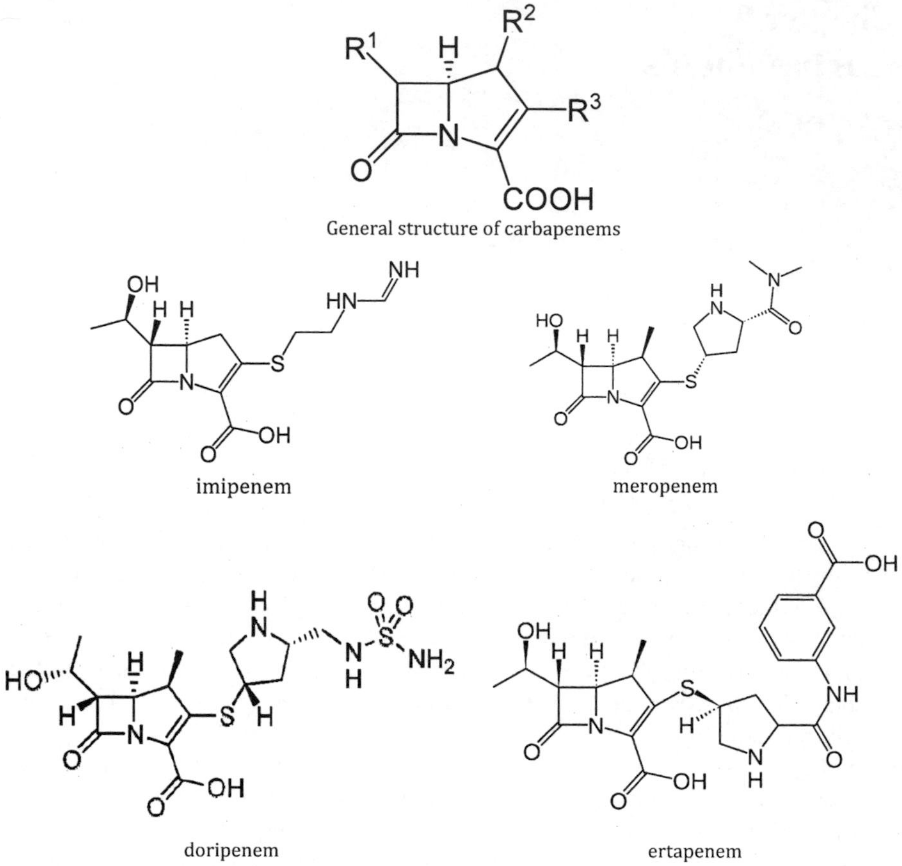

General structure of carbapenems

imipenem meropenem

doripenem ertapenem

Fig. 7.1 Structures of the basic carbapenem backbone, and the carbapenems: imipenem, meropenem, doripenem, and ertapenem

Enterobacteriaceae, *Pseudomonas aeruginosa, Burkholderia cepacia, Stenotrophomonas maltophilia* and *Staphylococcus epidermidis*. Anaerobes, including *B. fragilis*, are highly susceptible. Ertapenem (similar to the other carbapenems) is active against aerobic gram-positive and anaerobic organisms with the exception of *Pseudomonas* and *Acinetobacter*.

Common Uses

Carbapenems are effective against urinary tract, intra-abdominal, lower respiratory tract, gynecologic, skin, soft tissue, bone and joint infections.

Adverse Effects

The most frequent adverse events include nausea, vomiting, diarrhea, rash, drug fever, infusion site complications and seizure. Hematologic abnormalities such as eosinophilia, positive Coombs' test, minor liver enzyme elevation, thrombocytopenia, and increased prothrombin time have been reported.

Risk factors for seizure development include history of seizure activity, lesions of central nervous system and renal impairment. Carbapenems are not recommended for treatment of meningitis.

Imipenem

Imipenem is combined with cilastatin to inhibit degradation of imipenem by renal tubular dihydropeptidase. Cilastatin has no antimicrobial activity. The other carbapenems in the class do not need the addition of cilastatin since they are not hydrolyzed by dihydropeptidase.

Meropenem

Meropenem's spectrum of activity is similar to imipenem. It is slightly more active against aerobic gram-negative bacilli and slightly less active against aerobic gram-positive cocci. Its toxicity profile is similar to imipenem except that it may be less likely to cause seizures (0.5% vs. 1.5% for imipenem).

Ertapenem

Ertapenem differs from other carbapenems because of a longer half-life allowing once-a-day administration. However, it has no activity against *P. aeruginosa* and *Acinetobacter* spp.

Doripenem

The spectrum of activity of doripenem is similar to imipenem and meropenem. Doripenem may have a slight advantage over other carbapenems for the treatment of *P. aeruginosa* infections since it has the lowest MIC_{90} within this class of agents. Its side effect profile is similar to that of meropenem.

Summary

Imipenem, meropenem and doripenem have similar spectrums of activity and adverse event profiles. Ertapenem has no activity against *P. aeruginosa* and *Acinetobacter* spp. Its longer half-life makes it attractive for use in intra-abdominal and pelvic infections. The activities of carbapenems against a broad array of pathogens make them suitable for treatment of serious infections in which the causative organism is likely to be resistant to other anti-infective agents. The carbapenems should not be used as first-line agents unless the infecting organism has a multi-drug resistant profile and is only sensitive to carbapenems.

Further Reading

Baughman, R. P. (2009). The use of carbapenems in the treatment of serious infections. J Intensive Care Med, 24(4), 230–241.

Driscoll, J. A., Brody, S. L., & Kollef, M. H. (2007). The epidemiology, pathogenesis and treatment of Pseudomonas aeruginosa infections. Drugs, 67(3), 351–368.

Giamarellou, H., & Kanellakopoulou, K. (2008). Current therapies for pseudomonas aeruginosa. Crit Care Clin, 24(2), 261–278, viii.

Mazzei, T. (2010). The pharmacokinetics and pharmacodynamics of the carbapanems: focus on doripenem. J Chemother, 22(4), 219–225.

Shah, P. M. (2008). Parenteral carbapenems. Clin Microbiol Infect, 14 Suppl 1, 175–180.

Zhanel, G. G., Wiebe, R., Dilay, L., Thomson, K., Rubinstein, E., Hoban, D. J., et al. (2007). Comparative review of the carbapenems. Drugs, 67(7), 1027–1052.

Chapter 8
Glycopeptides and Other Agents Used to Treat Gram-Positive Cocci

Introduction

Glycopeptide antibiotics are a class of agents, composed of glycosylated cyclic or polycyclic peptides. Clinically useful glycopeptides include vancomycin, teicoplanin, dalavancin, and telavancin.

Vancomycin

Vancomycin is a glycopeptide first isolated in 1956 from *Streptomyces orientalis*. It is active against gram-positive bacteria and its use has been widespread in recent years with the emergence of MRSA (Fig. 8.1).

Mechanism of Action

Vancomycin inhibits bacterial cell wall synthesis (see Fig. 4.2) and demonstrates concentration-independent bactericidal activity. It inhibits peptidoglycan polymerase and transpeptidation reactions by forming a complex with the terminal D-alanyl-D-alanine portion of the peptide precursor units.

R.W. Finberg and R. Guharoy, *Clinical Use of Anti-infective Agents: A Guide on How to Prescribe Drugs Used to Treat Infections*,
DOI 10.1007/978-1-4614-1068-3_8,

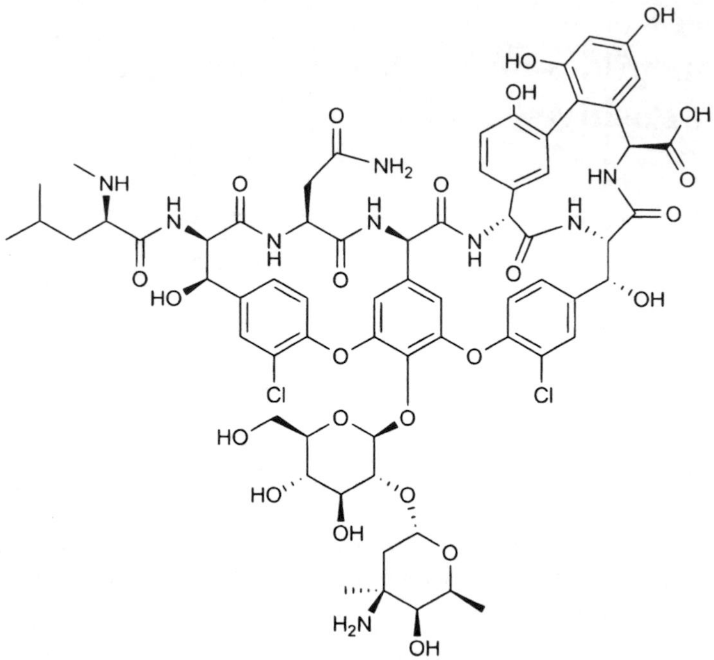

Fig. 8.1 Structure of vancomycin

Spectrum of Activity

The spectrum of activity of vancomycin includes all *Staphylococcus aureus*, coagulase-negative staphylococci, *Corynebacterium* spp., *Clostridium difficile*, *Enterococcus faecalis*, *Enterococcus faecium*, *Streptococcus pneumoniae*, and group A beta-hemolytic streptococci.

Common Uses

Vancomycin is the drug of choice for both coagulase-positive and -negative strains of *Staphylococcus aureus* infections including cellulitis, bacteremia, endocarditis, and osteomyelitis. It is also used for methicillin-sensitive *S. aureus* (MSSA) in patients allergic to penicillins or cephalosporins. It is the drug of choice for penicillin-resistant streptococci, *Corynebacterium* group *jeikeium*, penicillin-resistant enterococci, and *Bacillus* spp. The oral formulation is used for treatment of *Clostridium difficile*-associated severe infections. Overuse of vancomycin has created a challenging selection pressure reflected by the alarming emergence of vancomycin-resistant enterococci. The recent Infectious Diseases

Society of America (IDSA) therapeutic guidelines for the treatment of MRSA infections recommend the use of an alternative agent for MRSA isolates with a vancomycin MIC > 2 mcg/ml.

Adverse Effects

Rapid infusion of vancomycin results in non-immunologic histamine release associated "red man syndrome." Symptoms include erythema, pruritus, flushing of the upper torso, angioedema, and occasionally hypotension. Nephrotoxicity and ototoxicity are rare complications that can be potentiated by concomitant use of nephrotoxic agents. Vancomycin-induced neutropenia and thrombocytopenia are rare.

Teicoplanin

Teicoplanin is a glycopeptide with a similar spectrum of activity to vancomycin. It appears to be slightly more active in vitro than vancomycin against staphylococci and streptococci. However, it is less active against *Staphylococcus hemolyticus*. Adverse effects, such as local irritation at injection site and rash have been reported. Teicoplanin is not associated with histamine release which can contribute to "red man syndrome" as seen with vancomycin use. Teicoplanin is not commercially available in the US (Fig. 8.2).

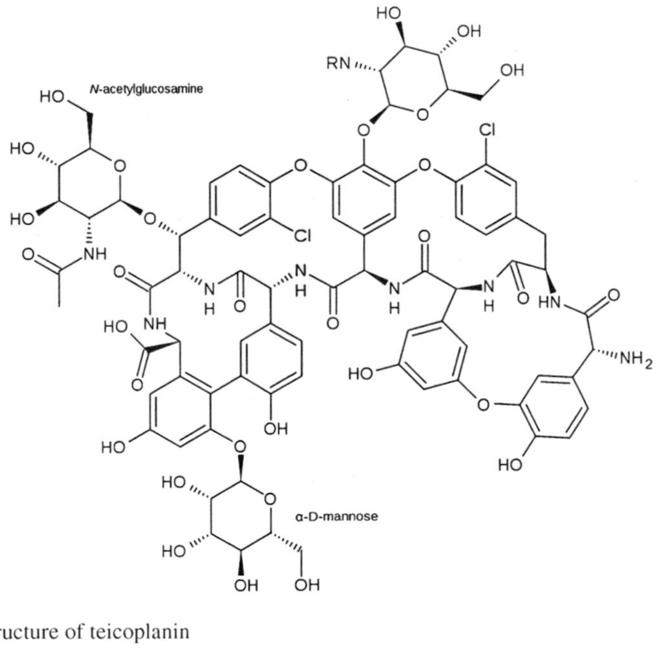

Fig. 8.2 Structure of teicoplanin

Dalbavancin and Telavancin

Dalbavancin and telavancin are structurally similar to vancomycin. Telavancin has been recently approved by the FDA for use in the US (Fig. 8.3). Dalbavancin is not currently approved for use in the US. Once approved, dalbavancin is expected to gain wider clinical use in the outpatient setting because its unique pharmacokinetic profile permits once a week administration.

Mechanism of Action

Like vancomycin, telavancin inhibits bacterial cell wall synthesis (see Fig. 4.2) and disrupts bacterial membranes by depolarization.

Spectrum of Activity

The spectrum of activity of telavancin includes MSSA and MRSA, *Streptococcus agalactiae*, *Streptococcus anginosus* group, and vancomycin-sensitive *Enterococcus faecalis*.

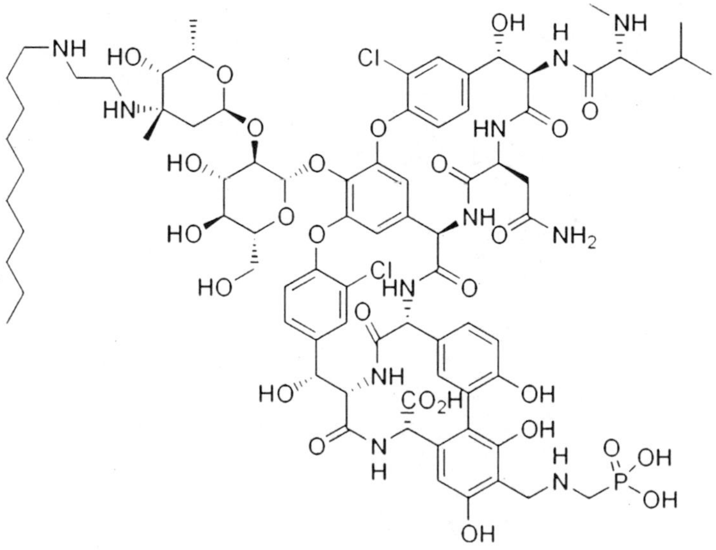

Fig. 8.3 Structure of telavancin hydrochloride

Common Uses

Televancin is approved for complicated skin and skin structure infections in adults. Use during pregnancy is not recommended because animal studies reported adverse effects in fetal development.

Adverse Effects

Televancin should be infused over 60 min to avoid vancomycin like "red man syndrome." Mild taste disturbance, nausea, vomiting, and mild renal impairment have been reported. Televancin has been reported to interfere with blood coagulation tests. Such tests should be performed 23-h after administration of telavancin. Occasional prolongation of the QT interval has also been reported.

Miscellaneous Non-glycopeptide Agents

Streptogramins

Streptogramins are a naturally occurring class of antibiotics isolated from *Streptomyces pristinaespiralis.*

Quinupristin/Dalfopristin

Quinupristin/dalfopristin (Q/D) is the only FDA-approved agent in this class in the US. Q/D is a synergistic combination of two streptogramins in a 30:70 ratio.

Mechanism of Action

Quinupristin and dalfopristin inhibit protein synthesis. They specifically bind to different sites on the 50S ribosome forming a stable tertiary complex that interferes with different targets of 23S RNA (Fig. 8.4). Dalfopristin inhibits the early phase and quinupristin inhibits the late phase of ribosome activity.

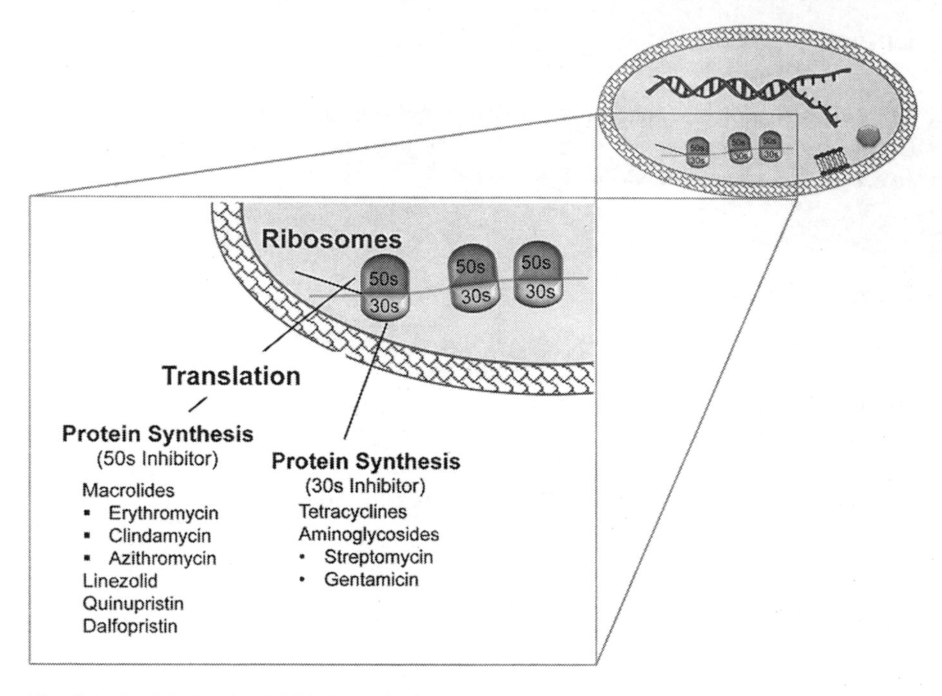

Fig. 8.4 Antibiotics that inhibit bacterial protein synthesis

Spectrum of Activity

The spectrum of activity for Q/D includes vancomycin-resistant *Enterococcus faecium*, MSSA, *Streptococcus pyogenes*, *Legionella* spp., *Fusobacterium*, and *Bacteroides* spp. It is not active against *Enterococcus faecalis.*

Adverse Effects

Side effects include venous irritation, elevation of conjugated bilirubin, headache, nausea, vomiting, diarrhea, and skin rash. Infusion via a central line is recommended to avoid venous irritation. Myalgias and arthralgias are severe troublesome side effects which are reversible after discontinuation of the agent. Q/D significantly inhibits cytochrome P450 3A4. Clinicians should be aware of concomitant use of drugs that are metabolized by P450 3A4 to avoid potential adverse events associated with drug interactions.

The availability of other therapeutic options, such as linezolid and daptomycin, has limited the use of Q/D (Fig. 8.5).

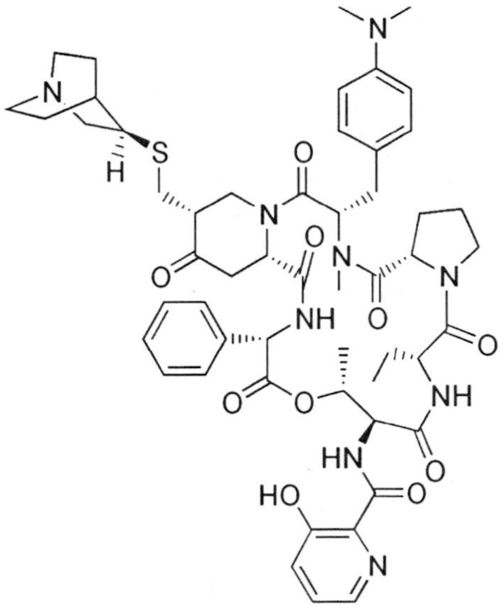

Fig. 8.5 Structure of quinupristin/dalfopristin

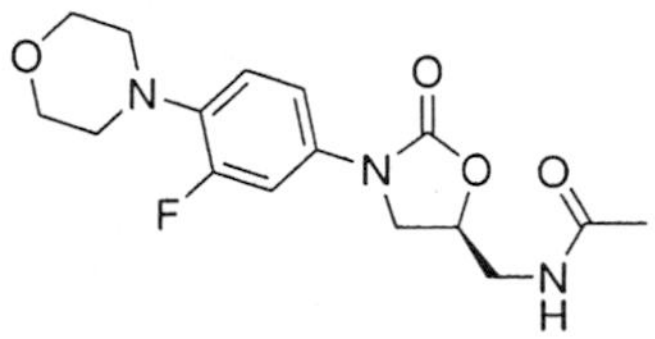

Fig. 8.6 Structure of linezolid

Oxazolidinones

The oxazolidinones are a new class of agents that have a unique mechanism of action against bacterial protein synthesis. Linezolid is the only oxazolidinone approved by the FDA for use in the US.

Linezolid

Linezolid is a morpholinyl analog of the piperazinyl oxazolidinone with a fluorine substitution at the phenyl-3 position for enhancement of antibacterial activity (Fig. 8.6).

Mechanism of Action

Linezolid inhibits bacterial protein synthesis by binding to the bacterial 23S ribosomal RNA of the 50S subunit and preventing the formation of the 60S initiation complex, an essential component of the bacterial translation process (see Fig. 8.4).

Spectrum of Activity

The spectrum of activity of linezolid includes vancomycin-resistant *Enterococcus faecium*, penicillin-resistant pneumococci, MSSA, MRSA, *Streptococcus pyogenes*, *Streptococcus agalactiae*, and *Streptococcus pneumoniae*.

Common Uses

Because of its 100% bioavailability and activity against MRSA, VRE and coagulase-negative staphylococci, linezolid use is increasing as a therapeutic option to treat pneumonia and skin infections caused by MRSA. It is indicated for the treatment of nosocomial pneumonia, complicated skin and skin structure infections caused by MSSA or MRSA infections. It is also indicated for uncomplicated skin and skin structure infections secondary to MSSA or *Streptococcus pyogenes*, and community-acquired pneumonia secondary to penicillin-sensitive *S. pneumoniae*.

Adverse Effects

Linezolid is a non-reversible monoamine oxidase inhibitor (MAOI). Its MAOI activity should be considered for potential drug interactions with adrenergic and serotonergic agents. Reversible time and dose-dependent myelosuppression, particularly thrombocytopenia, have been reported. Due to the potential for hypertensive crisis and hyperserotonemia as a consequence of the inhibition of dietary amine catabolism by MAOI, consumption of foods or beverages containing tyramine and tryptophan should be avoided.

Daptomycin

Daptomycin is a naturally occurring cyclic lipopeptide derived from *Streptomyces roseosporus* as a fermentation product. The structure consists of a ten-membered amino acid ring with a 10-carbon decanoic acid attached to a terminal L-tryptophan (Fig. 8.7).

Mechanism of Action

Daptomycin binds to the cell membrane in a calcium-dependent manner resulting in depolarization of the bacterial membrane and termination of bacterial DNA, RNA, and protein synthesis (see Fig. 3.2).

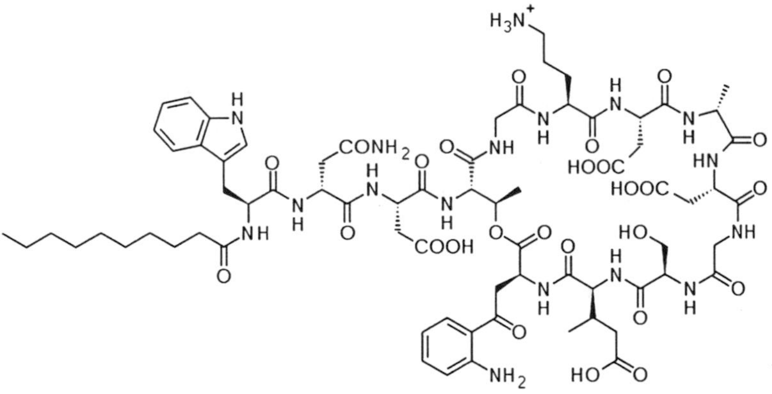

Fig. 8.7 Structure of daptomycin

Spectrum of Activity

The spectrum of activity of daptomycin includes MSSA, MRSA, streptococci, enterococcal species, and penicillin-resistant *Streptococcus pneumoniae*.

Common Uses

Daptomycin is approved for treatment of complicated skin and skin structure infections and bacteremia secondary to *Staphylococcus aureus*, *S. pyogenes*, *S. agalactiae*, and *Enterococcus faecalis*. Daptomycin should not be used for treatment of pneumonia since it binds to surfactant and this interaction inhibits the activity of daptomycin.

Adverse Effects

Elevation of creatine phosphokinase (CPK) and myopathy are the two major reversible toxicities of daptomycin. Weekly monitoring of CPK during therapy and discontinuation of therapy if CPK is elevated 5 times the normal is recommended. Other adverse effects include constipation, nausea, vomiting, diarrhea, dizziness, dyspnea, hypotension, and hypertension. Concurrent use with HMG-CoA reductase inhibitors may increase the risk of myopathy.

Fidaxomicin

Fidaxomicin is a narrow spectrum 18-membered macrocyclic oral agent approved by the FDA in May 2011. It offers a new treatment option for *Clostridium difficile* diarrhea. It is a member of the tiacumicin family and was isolated from *Dactylosporangium aurantiacum* (Fig. 8.8).

Mechanism of Action

Fidaxomicin works by inhibiting bacterial protein synthesis via inhibition of RNA polymerases. Like oral vancomycin, it has minimal systemic absorption. It is hydrolyzed to a microbiologically active compound and its metabolism is not dependent on cytochrome P450 (CYP) enzymes. More than 92% of fidaxomicin is excreted via feces and only 0.59% is excreted via urine. The elimination half-life of the metabolite is 11.2 ± 2.01 h.

Spectrum of Activity

In vitro, fidaxomicin demonstrated activity against *Clostridium* species, including *Clostridium difficile*. Published studies comparing in vitro activity of fidaxomicin with vancomycin and metronidazole, the traditional agents used for treatment of *C. difficile* infection, have demonstrated good activity against all strains of *C. difficile* strains tested. However, minimum inhibitory concentrations were consistently lower for fidaxomicin (0.03–0.25 mcg/ml).

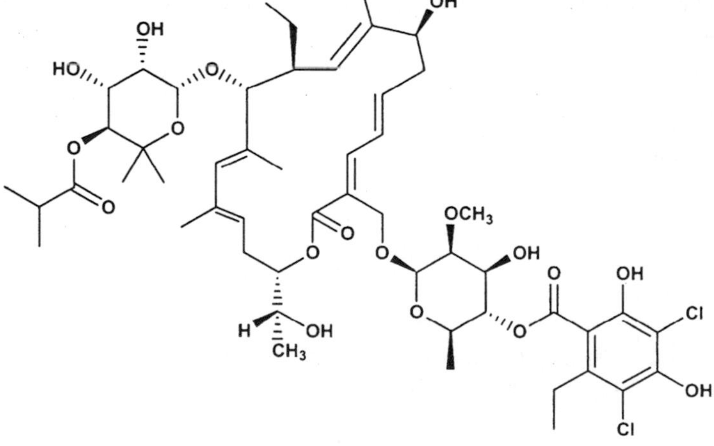

Fig. 8.8 Structure of fidaxomicin

Common Uses

It is approved by the FDA for use in adults (>18 years) for treatment of *C. difficile*-associated diarrhea. The recommended dose is 200 mg orally twice daily and can be administered with or without food. No dosage adjustment is necessary. In clinical studies, fidaxomicin was non-inferior to oral vancomycin and superior in reducing acute symptoms. However, one study reported no significant difference between fidaxomicin and vancomycin in the rate of recurrence in patients infected with the hypervirulent baseline *C. difficile* strains.

Adverse Effects

Common adverse effects reported in the clinical trials include nausea (11%), vomiting (7%), abdominal pain (6%), and gastrointestinal hemorrhage (4%). Other reported events (<2%) include pruritus, rash, hyperglycemia, metabolic acidosis, dysphagia, bowel obstruction, flatulence, elevated liver enzymes, neutropenia, and anemia. Teratogenicity or breast feeding effects in pregnant patients are unknown.

Fidaxomicin and the active metabolite are substrates of the efflux transporter P-glycoprotein. In vivo studies did not demonstrate the need for dosage adjustment when co-administered with another P-glycoprotein substrate.

Further Reading

Arias, C. A., Contreras, G. A., & Murray, B. E. (2010). Management of Multi-Drug Resistant Enterococcal Infections. Clin Microbiol Infect.

Credito, K. L., & Appelbaum, P. C. (2004). Activity of OPT-80, a novel macrocycle, compared with those of eight other agents against selected anaerobic species. Antimicrob Agents Chemother, 48(11), 4430–4434.

Curcio, D. (2010). Resistant pathogen-associated skin and skin-structure infections: antibiotic options. Expert Rev Anti Infect Ther, 8(9), 1019–1036.

Dificid® oral tablets [package insert]. San Diego, CA: Optimer Pharmaceuticals Inc, April, 2011.

Di Paolo, A., Malacarne, P., Guidotti, E., Danesi, R., & Del Tacca, M. (2010). Pharmacological issues of linezolid: an updated critical review. Clin Pharmacokinet, 49(7), 439–447.

Guskey, M. T., & Tsuji, B. T. (2010). A comparative review of the lipoglycopeptides: oritavancin, dalbavancin, and telavancin. Pharmacotherapy, 30(1), 80–94.

Kollef, M. H. (2009). New antimicrobial agents for methicillin-resistant Staphylococcus aureus. Crit Care Resusc, 11(4), 282–286.

Liu, C., Bayer, A., Cosgrove, S. E., Daum, R. S., Fridkin, S. K., Gorwitz, R. J., et al. (2011). Clinical practice guidelines by the infectious diseases society of america for the treatment of methicillin-resistant Staphylococcus aureus infections in adults and children: executive summary. Clin Infect Dis, 52(3), 285–292.

Louie, T. J., Miller, M. A., Mullane, K. M., Weiss, K., Lentnek, A., Golan, Y., et al. (2011). Fidaxomicin versus vancomycin for Clostridium difficile infection. N Engl J Med, 364(5), 422–431.

Nailor, M. D., & Sobel, J. D. (2009). Antibiotics for gram-positive bacterial infections: vancomycin, teicoplanin, quinupristin/dalfopristin, oxazolidinones, daptomycin, dalbavancin, and telavancin. Infect Dis Clin North Am, 23(4), 965–982, ix.

Pfeiffer, R. R. (1981). Structural features of vancomycin. Rev Infect Dis, 3 suppl, S205-209.

Rybak, M. J. (2006). The efficacy and safety of daptomycin: first in a new class of antibiotics for Gram-positive bacteria. Clin Microbiol Infect, 12 Suppl 1, 24–32.

Saravolatz, L. D., Stein, G. E., & Johnson, L. B. (2009). Telavancin: a novel lipoglycopeptide. Clin Infect Dis, 49(12), 1908–1914.

Sullivan, K. M., & Spooner, L. M. (2010). Fidaxomicin: a macrocyclic antibiotic for the management of Clostridium difficile infection. Ann Pharmacother, 44(2), 352–359.

Welte, T., & Pletz, M. W. (2010). Antimicrobial treatment of nosocomial methicillin-resistant Staphylococcus aureus (MRSA) pneumonia: current and future options. Int J Antimicrob Agents, 36(5), 391–400.

Chapter 9
Aminoglycosides

Introduction

Aminoglycosides are derivatives of naturally occurring organisms and several aminoglycosides have been introduced for clinical use since the discovery of streptomycin in 1947. Amikacin, gentamicin, kanamycin, netilmicin, neomycin, streptomycin, paromomycin, and tobramycin are currently approved by the FDA for use in the US. Among these, amikacin, gentamicin, and tobramycin are most frequently used. Because of the availability of less toxic agents, aminoglycosides have been used less often in recent years.

Mechanism of Action

Aminoglycosides are complex sugars bound by a glycosidic linkage to a central hexose nucleus (Fig. 9.1). They bind to the aminoacyl site of the 16S ribosomal RNA within the 30S ribosomal subunit. This binding interferes with reading of the genetic code resulting in inhibition of protein synthesis (see Fig. 8.4).

Spectrum of Activity

The spectrum of activity of aminoglycosides includes a wide range of aerobic gram-negative bacilli, including *Enterobacteriaceae*, *Pseudomonas* spp., *Haemophilus influenzae*, many staphylococci, and specific mycobacteria. They are also active against MSSA. Activities against *Burkholderia cepacia*, *Stenotrophomonas maltophilia*, and anaerobic bacteria are usually poor or absent.

R.W. Finberg and R. Guharoy, *Clinical Use of Anti-infective Agents: A Guide on How to Prescribe Drugs Used to Treat Infections*,
DOI 10.1007/978-1-4614-1068-3_9,

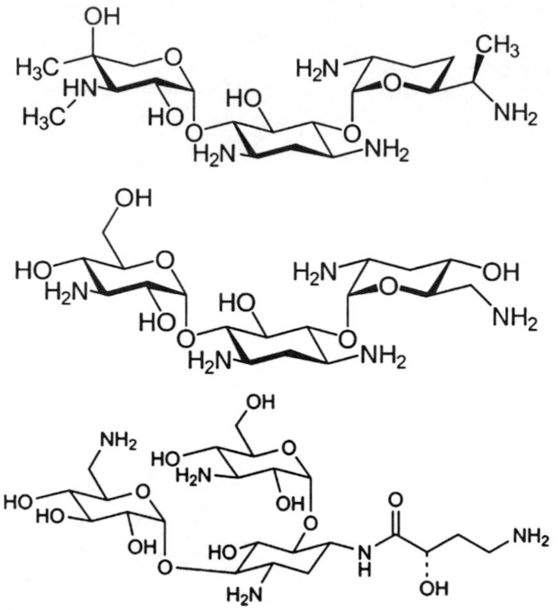

Gentamicin

Tobramycin

Amikacin

Fig. 9.1 Structures of gentamicin, tobramycin, and amikacin

Pharmacokinetics

Aminoglycosides are poorly absorbed from the gastrointestinal tract. They are highly polar cations and their distribution is usually limited to the extracellular fluid compartment. They are primarily excreted by the kidney via glomerular filtration. The serum half-life is 2–3 h in patients with normal kidney function and prolonged in patients with impaired renal function. They are removed effectively by continuous hemofiltration or hemodialysis.

Traditional Multiple Daily Dosing

The recommended traditional dosing begins with the administration of a "loading dose" (1.5–2 mg/kg for gentamicin, tobramycin and 7.5–15 mg/kg for amikacin). Loading dose is based on ideal body weight (IBW):

- IBW for male patients = 50 kg + 2.3 kg (height in inches > 60 inches)
- IBW for female patients = 45 kg + 2.3 kg (height in inches > 60 inches)

Table 9.1 Nomogram of creatinine

Creatinine clearance (ml/min)	Maintenance dose (% of loading dose)	Dose interval (h)
>90	84	8
80	80	8
70	76	8
60	84	12
50	79	12
40	72	12
30	86	24
20	75	24–36
<20[a]		

[a]If CrCl < 20 ml/min: a loading dose recommended with subsequent doses based on serum aminoglycoside concentrations

- Dosing weight for obese patients (>30% IBW) is calculated based on reduced volume of distribution in the adipose tissue: Dosing weight = IBW + 0.4 (actual weight in kg – IBW)

The estimated creatinine clearance is calculated by using the Cockcroft-Gault equation:

- Creatinine clearance (ml/min) for male patients = [(140–age) × Weight (kg)]/[serum creatinine × 72]
- Creatinine clearance for female patients = [(140–age) × Weight (kg)]/[serum creatinine × 72] × 0.85

Maintenance dose can be calculated by using the nomogram of creatinine (Table 9.1).

Monitoring of serum concentrations has been the standard of care in patients receiving traditional dosing regimens. Accuracy in administration and sampling time is critical; peak concentrations should be determined 30 min after completion of a 30-min intravenous infusion and trough concentrations should be measured immediately before the next dose is administered. Peak concentrations are dependent upon the indication and site of infection. For example, peak levels for gentamicin and tobramycin should be 3–4 mcg/ml when given for synergy (e.g., gram-positive infections) and 6–9 mcg/ml for serious invasive infections.

Single Daily Dose

A once-daily dose regimen has two advantages over the traditional dosing method and offers comparable efficacy with the possibility of decreased nephrotoxicity and ease of administration. The pharmacodynamic properties of aminoglycosides provide the basis of the advantages and they are based on post-antibiotic effect (PAE) and concentration-dependent killing. PAE achieves persistent inhibitory or cidal effects against gram-negative organisms after the drug is eliminated from the body.

Table 9.2 Determination of aminoglycoside dosing

Creatinine clearance (ml/min)	Initial and maintenance dose (mg)	Dosing interval (h)
>60	7 mg/kg	24
40–49	7 mg/kg	36
<40	Use traditional method	

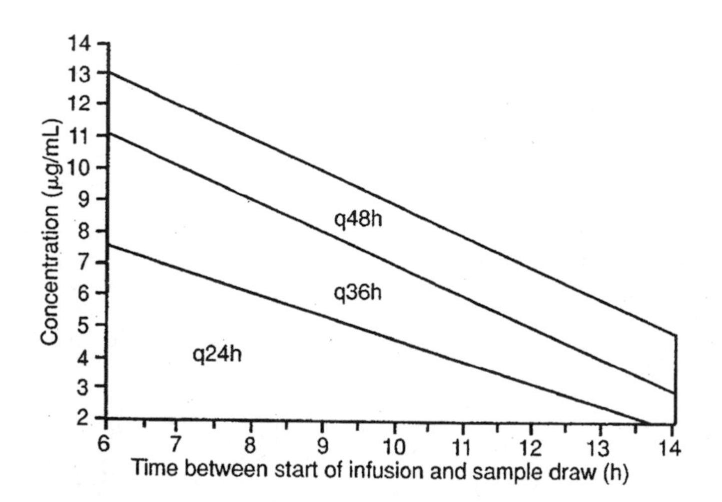

Fig. 9.2 Single daily dose aminoglycoside nomogram

These effects usually last between 1 and 7.5 h post administration. The other property, "concentration-dependent killing," allows the administration of high concentrations of the drug resulting in rapid killing of the organism. It is believed that once-daily dosing potentially achieves maximum peak concentrations via concentration-dependent killing. Animal studies have suggested reduction of incidence of renal failure with the once-daily dosing method. However, there is no clear evidence of renal protection with this type of dosing in humans. Single daily dosing is of limited value in patient populations with altered pharmacokinetics, such as cystic fibrosis patients, critically ill patients, and patients with fluctuating serum creatinine concentrations.

Initial dosing utilizes a dose of gentamicin or tobramycin 7 mg/kg with an interval of 24–48 h. The process starts with calculation of IBW as described above. Then, estimated creatinine clearance is calculated using the Cockcroft-Gault equation, also described above.

The dosing interval is calculated based upon the estimated creatinine clearance (Table 9.2).

A single serum concentration is obtained between 6 and 14 h after the first dose and the nomogram shown in Fig. 9.2 is used to adjust the necessary dosing interval.

Adverse Effects

The primary toxicities are nephrotoxicity and ototoxicity. The incidence of nephrotoxicity depends upon the specific patient population and concomitant risk factors. The incidence has been reported as high as 20%. The typical onset occurs after a treatment of 6–10 days. Overt ototoxicity generally occurs in 2–10% of patients. Vestibular or cochlear damage is manifested as ototoxicity. Vestibular toxicity may include lightheadedness, vertigo, ataxia, nausea and vomiting. Therefore, careful monitoring is crucial to avoid potential adverse events. In most cases, the toxicities are reversible. Neuromuscular blockade, secondary to concomitant drug therapy or disease states, can occur on rare occasions.

Common Uses

The most common uses of aminoglycosides include empiric therapy for nosocomial respiratory tract infections, complicated urinary tract infections, sepsis, intra-abdominal infections, and osteomyelitis secondary to aerobic gram-negative bacilli. Aminoglycosides are frequently used in combination with a beta-lactam agent. Because of the potential toxicities of aminoglycosides, therapy is streamlined once an organism is identified in most situations.

Streptomycin can be used for endocarditis secondary to *E. faecalis* or viridans streptococci in strains that are susceptible to high levels of streptomycin (use with penicillin or vancomycin). Streptomycin is also used to treat plague, tularemia, refractory active tuberculosis, and acute brucellosis (in combination with tetracycline). It has less nephrotoxicity and greater vestibular toxicity when compared with gentamicin.

Neomycin, kanamycin, and paromomycin are not utilized for systemic use because of toxicities. These three agents are closely related with similar spectrum and cross-resistance. Neomycin is used for preparation of bowel surgery with 1 g dose administered orally every 6–8 h 1 day prior to surgery (with erythromycin) to reduce bowel flora. Paromomycin is poorly absorbed and its use is limited to treat asymptomatic intestinal amebiasis and giardiasis in pregnancy. Amikacin is used to treat serious gram-negative infections including many strains of *Pseudomonas*, *Enterobacter*, *Serratia*, and *Proteus*. It is resistant to many enzymes that inactivate gentamicin and tobramycin. Multi-drug-resistant *Mycobacterium tuberculosis* is usually susceptible to amikacin.

Gentamicin and tobramycin have similar spectrums with the exception that gentamicin is slightly more active against *Serratia* and tobramycin slightly more active against *Pseudomonas*. They are mainly used for severe infections such as pneumonia and sepsis secondary to gram-negative infections.

Further Reading

Nicolau, D., Quintiliani, R., & Nightingale, C. H. (1992). Once-daily aminoglycosides. Conn Med, 56(10), 561–563.

Nicolau, D. P., Freeman, C. D., Belliveau, P. P., Nightingale, C. H., Ross, J. W., & Quintiliani, R. (1995). Experience with a once-daily aminoglycoside program administered to 2,184 adult patients. Antimicrob Agents Chemother, 39(3), 650–655.

Siber, G. R., Echeverria, P., Smith, A. L., Paisley, J. W., & Smith, D. H. (1975). Pharmacokinetics of gentamicin in children and adults. J Infect Dis, 132(6), 637–651.

Zaske, D. E., Cipolle, R. J., Rotschafer, J. C., Solem, L. D., Mosier, N. R., & Strate, R. G. (1982). Gentamicin pharmacokinetics in 1,640 patients: method for control of serum concentrations. Antimicrob Agents Chemother, 21(3), 407–411.

Chapter 10
Macrolides

Introduction

Macrolides (erythromycin, azithromycin, and clarithromycin) are a group of antibiotics with a macrocyclic lactone nucleus attached to sugar moieties (Fig. 10.1). They are derived from *Streptomyces erythreus*. Azithromycin and clarithromycin are derived from structural modifications of erythromycin and result in improved absorption, dosing, spectrum of activity, and administration. Clarithromycin has the same 14-membered lactone ring as erythromycin replacing the hydroxyl group by a methoxy-group at position six. Azithromycin is a 15-membered lactone ring replacing the carbonyl group of erythromycin by methyl-substituted nitrogen at position 9A.

Mechanism of Action

These drugs bind specifically to the 50S portion of the bacterial ribosome and inhibit bacterial protein synthesis (see Fig. 8.4).

Spectrum of Activity

Erythromycin

Erythromycin is active against group A beta-hemolytic streptococci and penicillin-susceptible *Streptococcus pneumoniae* (penicillin-resistant strains are an exception). The majority of the strains of groups B, C, F, and G streptococci are susceptible to erythromycin. Erythromycin is active against *Neisseria gonorrhea, N. meningitidis, Bordetella pertussis, Treponema pallidum, Mycoplasma pneumoniae,*

R.W. Finberg and R. Guharoy, *Clinical Use of Anti-infective Agents: A Guide on How to Prescribe Drugs Used to Treat Infections*,
DOI 10.1007/978-1-4614-1068-3_10,

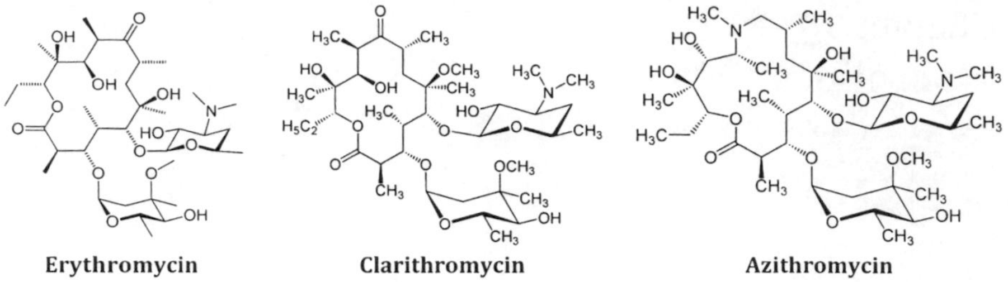

Erythromycin Clarithromycin Azithromycin

Fig. 10.1 Structures of Macrolides: erythromycin, clarithromycin, and azithromycin

Chlamydophila species, *Mycoplasma catarrhalis*, *Ureaplasma urealyticum*, *Legionella pneumophila*, some strains of *Rickettsia*, and MSSA. Erythromycin has no activity against *Mycobacterium avium* complex (MAC), whereas both azithromycin and clarithromycin have activity against MAC. In addition, both azithromycin and clarithromycin provide additional gram-negative activity compared with erythromycin including *Escherichia coli*, *Salmonella* spp., *Shigella* spp., *Campylobacter jejuni*, and *Helicobacter pylori*.

Erythromycin is excreted primarily in the bile; only 2–5% is excreted in the urine.

Azithromycin

The spectrum of activity of azithromycin is much broader than erythromycin and includes *E. coli*, *Shigella*, *Salmonella*, and *H. pylori*. *H. influenzae* is susceptible to azithromycin, whereas it is partially resistant to clarithromycin and completely resistant to erythromycin. It appears that azithromycin is most active against *Legionella* species compared with erythromycin and clarithromycin.

Azithromycin is excreted in the bile and dosage adjustment is not necessary for patients with impaired renal function.

Clarithromycin

The spectrum of activity of clarithromycin is similar to erythromycin and includes some strains of *H. influenzae*. It provides the best activity against MAC compared with azithromycin. However, drug–drug interactions are less common with azithromycin.

Clarithromycin is extensively metabolized in the liver, utilizing the cytochrome P450 system. The major active metabolite is 14-hydroxy-clarithromycin and 20–30% of clarithromycin is excreted unchanged in the urine. The dose should be reduced by half or the dosing interval should be doubled in patients with creatinine clearances below 30 ml/min.

Resistance

The primary means of resistance occurs by post-transcriptional methylation of the 23S bacterial ribosomal RNA. Cross-resistance with macrolides resulting in resistance to MRSA and enterococci has been reported.

Common Uses

Macrolides can be used to treat a wide variety of upper and lower respiratory tract infections including community-acquired pneumonia, soft tissue infections, pharyngitis, acute bronchitis, and sinusitis. Clarithromycin is used to treat *H. pylori* and in both prophylaxis and treatment of MAC infection.

Macrolides are used as an alternative for patients with penicillin allergy.

Adverse Effects

The most common side effects associated with erythromycin are nausea, vomiting, abdominal pain, and diarrhea. Transient hearing loss and QT prolongation have been reported.

Both azithromycin and clarithromycin are better tolerated than erythromycin and reported to cause less gastrointestinal side effects. Mild and transient elevations of liver enzymes and leucopenia have been reported. Higher dosage regimens (1,000 mg twice a day) used to treat mycobacterial infections are associated with higher incidence of gastrointestinal side effects. QT interval prolongation has been reported with both erythromycin and clarithromycin, but azithromycin has a minimal effect on QT prolongation.

Erythromycin and clarithromycin inhibit the metabolism of many drugs via interference with the hepatic P450 enzyme system, particularly CYP3A4. This enzyme interaction may result in elevation of serum levels of carbamazepine, theophylline, warfarin, triazolam, alfentanil, bromocriptine, cyclosporine, ritonavir, cimetidine, colchicine, digoxin, ergot alkaloids, lovastatin, simvastatin, phenytoin, pimozide, rifampin, rifabutin, tacrolimus, and valproic acid. On the other hand, there are few interactions with azithromycin since it does not appear to affect hepatic enzymes significantly. Cyclosporine, digoxin, and pimozide serum levels may be increased with the concurrent use of azithromycin.

Further Reading

Alvarez-Elcoro, S., & Enzler, M. J. (1999). The macrolides: erythromycin, clarithromycin, and azithromycin. *Mayo Clin Proc, 74*(6), 613–634.

Griffith, R. S. (1986). Pharmacology of erythromycin in adults. *Pediatr Infect Dis, 5*(1), 130–140.

Peters, D. H., & Clissold, S. P. (1992). Clarithromycin. A review of its antimicrobial activity, pharmacokinetic properties and therapeutic potential. *Drugs, 44*(1), 117–164.

Chapter 11
Metronidazole and Clindamycin

Introduction

Metronidazole compound (1-β-hydroxyethyl-2-methyl-5-nitroimidazole) is a commonly used agent for the treatment of anaerobic infections. It is the only one of the five nitroimidazoles (nimorazole, carnidazole, metronidazole, sulnidazole, and tinidazole) approved by the FDA for use in the US (Fig. 11.1).

Mechanism of Action

Metronidazole works by an immediate inhibition of DNA synthesis. The low molecular weight compound freely diffuses across cell membranes and it also penetrates cells via porins. The mechanism of action involves intracellular reduction of the nitro group by susceptible organisms to form free radicals resulting in DNA damage and inhibiting enzymes involved in energy production (see Fig. 2.1).

Spectrum of Activity

Metronidazole is active against a wide variety of anaerobic bacteria and protozoal parasites. It is also active against *Trichomonas*, *Entamoeba histolytica*, and *Giardia lamblia*. It manifests activity against all anaerobic cocci and anaerobic gram-negative bacilli, including *Bacteroides* spp., *Clostridium* spp., *Prevotella* spp., *Fusobacterium* spp., and anaerobic spore forming gram-positive bacilli. Facultative anaerobic bacteria, aerobic and nonspore-forming gram-positive bacilli are usually resistant. Resistant organisms include *Proprionibacterium* spp., many species of *Bifidobacterium*, *Actinomyces,* and *Arachnia*.

R.W. Finberg and R. Guharoy, *Clinical Use of Anti-infective Agents: A Guide on How to Prescribe Drugs Used to Treat Infections*,
DOI 10.1007/978-1-4614-1068-3_11,

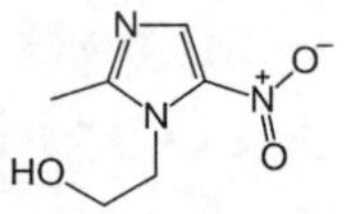

Fig. 11.1 Structure of metronidazole

Common Uses

Metronidazole is used to treat genital infections with *Trichomonas vaginalis* with cure rates of more than 90% of cases. However, resistant strains are increasingly reported in the literature. It is the drug of choice for all forms of amebiasis. Metronidazole is used for the treatment of anaerobic and microaerophilic organisms, including *Bacteroides*, *Clostridium*, *Fusobacterium*, *Peptococcus*, *Peptostreptococcus*, *Eubacterium*, and *Helicobacter*. It is frequently used as a first-line therapy for *Clostridium difficile* infections.

Adverse Effects

Adverse effects are usually not severe enough to require discontinuation of therapy. Commonly observed effects include nausea, headache, dry mouth, and a metallic taste. Other occasional effects include abdominal distress, vomiting, and diarrhea. Neurotoxic adverse effects, such as ataxia, dizziness, vertigo, encephalopathy, and convulsion have been reported and, if present, therapy should be discontinued. Therapy should also be discontinued in the event of numbness or paresthesias of extremities.

Clinicians should be well aware of the disulfiram-like effect which occurs with consumption of alcohol during or within 3 days of therapy. The typical symptoms include abdominal distress, headache, flushing, and vomiting. Patients should be counseled to avoid alcohol during therapy.

Clindamycin

Clindamycin, a derivative of lincomycin, was first isolated from *Streptomyces lincolnesis* in 1962 and became commercially available in 1966 (Fig. 11.2). It replaced lincomycin use because of its better absorption and clinical spectrum. It is active against gram-positive, gram-negative, and anaerobic organisms.

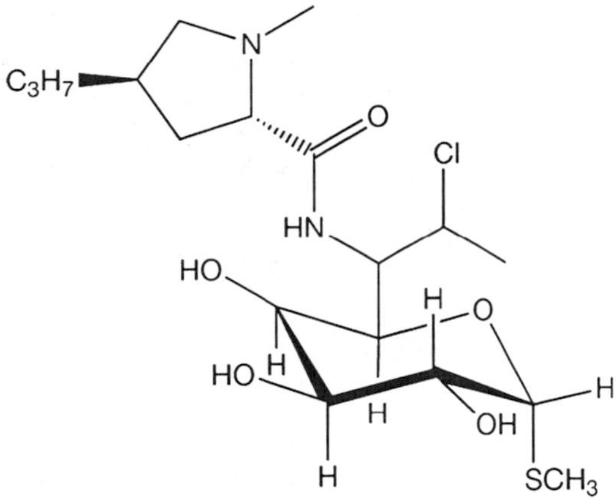

Fig. 11.2 Structure of clindamycin

Mechanism of Action

Clindamycin inhibits bacterial protein synthesis by binding to the 50S ribosomal subunit. The use of clindamycin with macrolides is not recommended since both of them compete for binding sites to the 50S subunit (see Fig. 8.4).

Spectrum of Activity

Clindamycin is active against most anaerobes including most anaerobic gram-positive cocci, nonspore forming bacilli, *Clostridium* (except *C. difficile* and some non-perfringens species), anaerobic gram-negative bacilli (except *Fusobacterium varium*), and anaerobic gram-positive nonspore forming bacilli.

It is active against most aerobic gram-positive organisms, including viridans streptococci, most strains of pneumococci, and other strains of streptococcal organisms (except *Enterococcus*). It is active against MSSA.

Common Uses

Clindamycin is effective in the treatment of most infections secondary to anaerobes and gram-positive cocci. It can be used for anaerobic pulmonary, intra-abdominal, gynecologic, pelvic, diabetic foot, and decubitus ulcer infections. Another appropriate agent should be added since the majority of these infections are polymicrobial. It can also be used as an alternative agent for patients with severe penicillin allergy. It is also used to treat *Clostridium perfringens* infection.

Oral preparations of clindamycin and vaginal cream are alternatives to metronidazole for the treatment of bacterial vaginosis. Topical solution is used for treatment of acne vulgaris and rosacea.

Clindamycin is extensively metabolized by the liver and the half-life is prolonged in patients with cirrhosis and hepatitis. Dose reductions are recommended in patients with acute liver disease.

Adverse Effects

The most commonly observed adverse effect is diarrhea. The reported incidence of *C. difficile* colitis in patients treated with clindamycin varies from 0.1 to 10%. The syndrome may be fatal. If the patient develops *C. difficile* colitis, clindamycin should be discontinued and the patient should be treated for *C. difficile*. Other side effects include rash, nausea, vomiting, diarrhea, flatulence, abdominal distension, anorexia, and transient elevation of liver enzymes. Other less common events, such as fever, neutropenia, thrombocytopenia, and eosinophilia have been reported.

Further Reading

Dhawan, V. K., & Thadepalli, H. (1982). Clindamycin: a review of fifteen years of experience. Rev Infect Dis, 4(6), 1133–1153.

Falagas, M. E., & Gorbach, S. L. (1995). Clindamycin and metronidazole. Med Clin North Am, 79(4), 845–867.

Kasten, M. J. (1999). Clindamycin, metronidazole, and chloramphenicol. Mayo Clin Proc, 74(8), 825–833.

Wexler, H. M. (2007). Bacteroides: the good, the bad, and the nitty-gritty. Clin Microbiol Rev, 20(4), 593–621.

Chapter 12
Quinolones

Introduction

Nalidixic acid was the first quinolone introduced in 1962. Structural changes have resulted in an extended spectrum and better pharmacokinetics profile. All agents have an N-1-substituted 1,4-dihydro-4-oxopyridine-3-carboxylic acid as the basic nucleus. The fluorinated 4-quinolones, such as norfloxacin, ciprofloxacin, moxifloxacin, and levofloxacin are effective for treatment of a wide variety of infectious diseases (Fig. 12.1). Potentially life threatening and rare side effects detected by postmarketing surveillance resulted in the withdrawal of gatifloxacin (hypoglycemia), trovafloxacin (hepatotoxicity), sparfloxacin (QTc prolongation, phototoxicity), temafloxacin (hemolytic anemia), grepafloxacin (cardiotoxicity), and clinafloxacin (phototoxicity) from the US market. Commercially available fluorinated quinolones available in the US are widely used by clinicians.

Mechanism of Action

Quinolones inhibit DNA synthesis via interference with DNA gyrase and the topoisomerase enzyme. DNA gyrase is the target for many gram-negative bacteria and topoisomerase for gram-positive organisms (Fig. 12.2).

Spectrum of Activity

First generation nonfluoroquinolones (nalidixic acid, cinoxacin, oxolinic acid) with limited antibacterial levels after oral administration are only useful for treatment of urinary tract infections. The fluoroquinolones with extended spectrum

R.W. Finberg and R. Guharoy, *Clinical Use of Anti-infective Agents: A Guide on How to Prescribe Drugs Used to Treat Infections*,
DOI 10.1007/978-1-4614-1068-3_12, © Springer Science+Business Media, LLC 2012

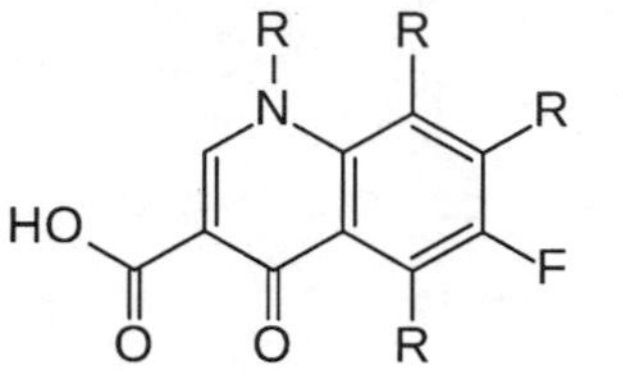

Basic quinolone structure

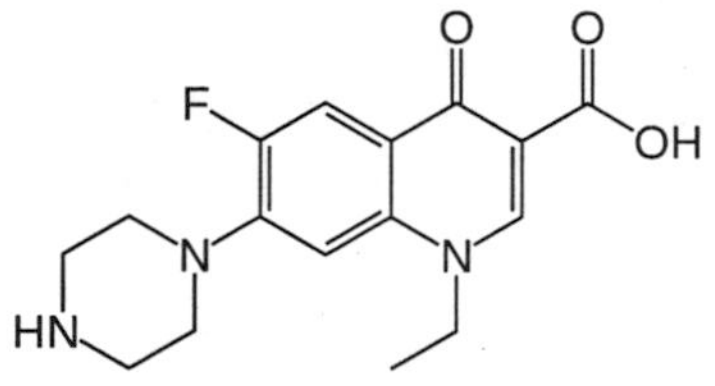

Norfloxacin

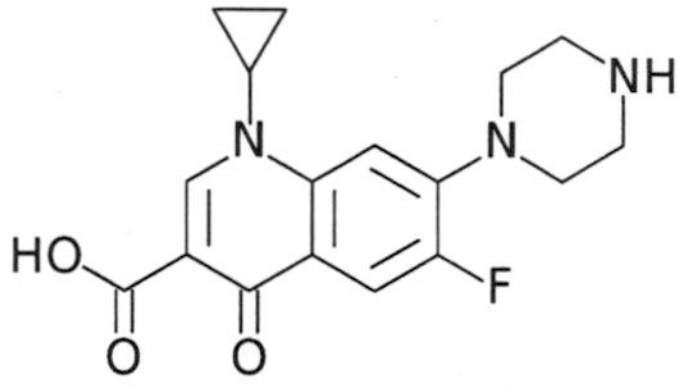

Ciprofloxacin

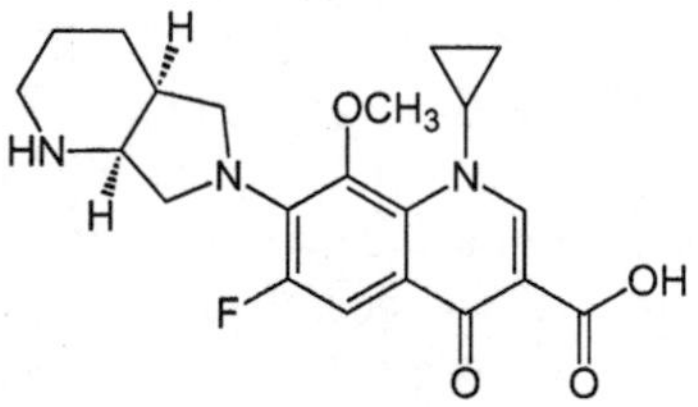

Moxifloxacin

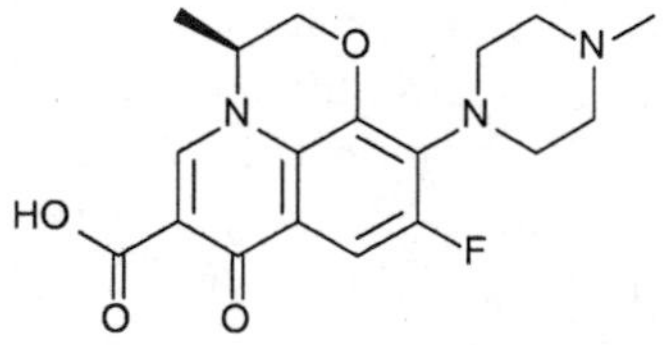

Levofloxacin

Fig. 12.1 Structures of the basic quinolone backbone, and the quinolones: norfloxacin, ciprofloxacin, moxifloxacin, and levofloxacin

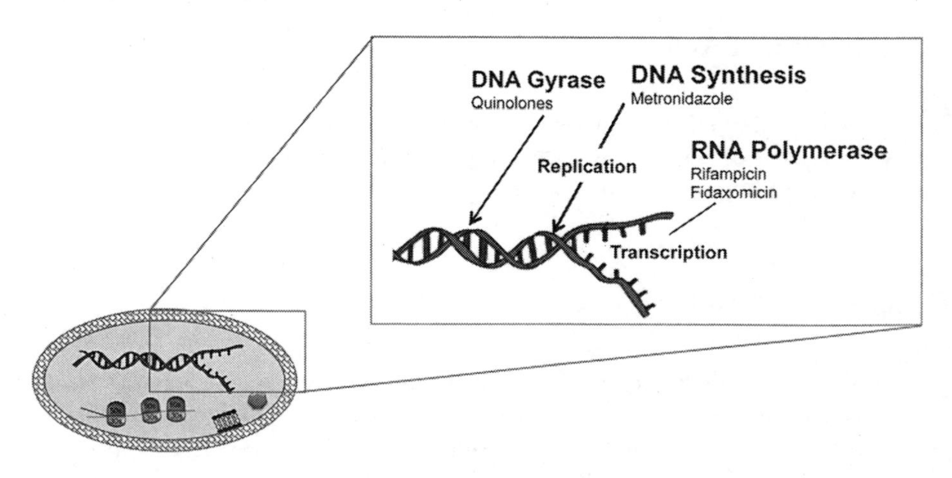

Fig. 12.2 Mechanism of action of antibiotics that block DNA and RNA replication

offer therapeutic serum and tissue levels. Fluoroquinolones are active against *Enterobacteriaceae*, as well as other gram-negative organisms such as *Haemophilus*, *Branhamella*, *Neisseria*, *Brucella*, *Salmonella*, *Legionella*, *Shigella*, *Campylobacter*, *Vibrio*, *Yersinia*, and *Aeromonas*. Ciprofloxacin and levofloxacin are the only quinolones effective against *Pseudomonas aeruginosa*. Quinolones are effective against *Mycobacterium tuberculosis*, *M. kansasii*, and *M. fortuitum*. Levofloxacin, moxifloxacin, and gemifloxacin have the best activity against gram-positive organisms, including *Staphylococcus aureus*, *S. epidermidis*, and *S. pneumoniae*. Indiscriminate use of quinolones has resulted in growing resistance against *E. coli*, *S. aureus*, and *P. aeruginosa*. Moxifloxacin provides modest coverage against some of the anaerobic pathogens, including *Bacteroides fragilis* and mouth anaerobes.

Common Uses

The fluoroquinolones are widely distributed in body fluids and tissues. They are excreted primarily in the urine requiring dosage adjustment in patients with renal impairment.

Fluoroquinolones are effective for treatment of prostatitis, enteritis, sexually transmitted diseases, urinary tract, lower respiratory tract, gynecologic, and soft tissue infections.

Patients should be warned of potential drug interactions with cations, nonsteroidal anti-inflammatory agents, and warfarin.

In the pediatric population, the use of fluoroquinolones is limited to treatment of lower respiratory infections in cystic fibrosis and life-endangering infections (see adverse effects below).

Adverse Effects

Common side effects include dizziness, mild nausea, vomiting, headache, and insomnia. Other less common adverse events include skin rashes, elevation of hepatic enzymes, seizure, hypoglycemia, leucopenia, and eosinophilia. QTc prolongation has been reported with moxifloxacin and caution should be exercised when used in patients on class III and class IA antiarrhythmic agents. Torsades de pointes have been reported with ciprofloxacin and levofloxacin. Quinolones are not commonly prescribed for children due to reports of joint and cartilage damage in juvenile animal studies. Tendinitis, including rupture of the Achilles tendon, has been seen in adults treated with quinolones.

Conclusion

Fluoroquinolones are one of the most useful classes of antimicrobial agents because of their extended spectrum and pharmacokinetic profile. Indiscriminate uses of these agents have resulted in increased resistance and clinicians should judiciously use these agents for appropriate indications.

Further Reading

Leibovitz, E. (2006). The use of fluoroquinolones in children. Curr Opin Pediatr, 18(1), 64–70.
O'Donnell, J. A., & Gelone, S. P. (2000). Fluoroquinolones. Infect Dis Clin North Am, 14(2), 489–513, xi.
O'Donnell, J. A., & Gelone, S. P. (2004). The newer fluoroquinolones. Infect Dis Clin North Am, 18(3), 691–716, x.
Walker, R. C. (1999). The fluoroquinolones. Mayo Clin Proc, 74(10), 1030–1037.

Chapter 13
Tetracyclines

Introduction

Tetracyclines are a class of antibiotics that function by inhibiting protein synthesis. Although they have a very broad spectrum of antibacterial activity, their current clinical use is primarily for the treatment of intracellular organisms like *Mycoplasma*, *Chlamydia*, and *Rickettsia*. They are also used as alternative agents for the treatment of syphilis in the case of penicillin allergy.

Tetracyclines are complex ring compounds consisting of a four-ring structure. The individual drugs are synthesized by additions to the common backbone (Fig. 13.1).

Tigecycline, an intravenously delivered broad-spectrum antibiotic (see below), is a glycylcycline. It is a semisynthetic derivative of minocycline (Fig. 13.2).

Mechanism of Action

Tetracyclines bind to the 30s subunit of the bacterial ribosome and block the binding of the aminoacyl-tRNA to the ribosome complex preventing the addition of peptides during protein synthesis (see Figs. 2.1 and 8.4). Tetracyclines are usually bacteriostatic drugs as they do not kill the bacteria but merely prevent their growth. Resistance to tetracyclines can occur in a variety of ways including impaired transport into the bacterium or enhanced transport out of the bacterium, alteration of the bacterial ribosomes, and enzyme inactivation.

R.W. Finberg and R. Guharoy, *Clinical Use of Anti-infective Agents: A Guide on How to Prescribe Drugs Used to Treat Infections*,
DOI 10.1007/978-1-4614-1068-3_13,

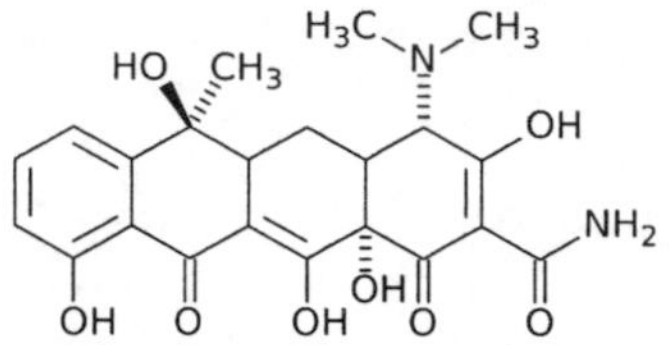

Fig. 13.1 Structure of tetracycline

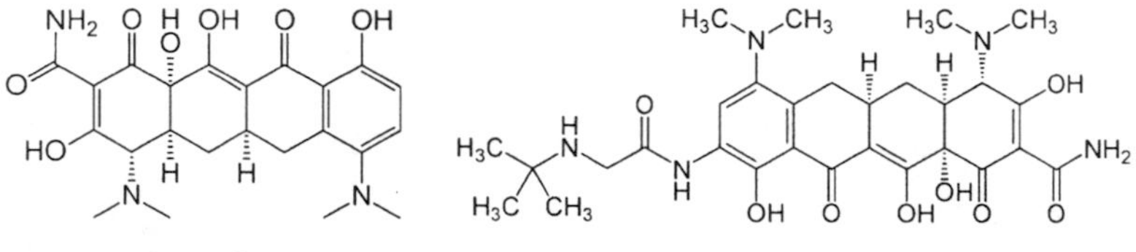

Minocycline

Tigecycline

Fig. 13.2 Structures of minocycline and tigecycline

Spectrum of Activity

Tetracyclines have activity against a broad spectrum of both gram-positive organisms (including *Streptococcus* and *Staphylococcus* spp.) as well as gram-negative organisms (not including *Pseudomonas*). They also have excellent activity against anaerobes. Not surprisingly, in view of their broad spectrum and reliable oral absorption, tetracyclines have been used for a large variety of indications. The development of resistance by *S. pneumoniae* has limited its usefulness in treating pneumonia. Similarly, although these agents have a broad spectrum of activity against gram-negative organisms, the development of resistance precludes its use as a first-line agent for urinary tract infections. The tetracyclines, principally doxycycline, are still widely used to treat intracellular pathogens, primarily rickettsial diseases but also *Chlamydia* and *Mycoplasma*.

Doxycycline is now the tetracycline most commonly used and the drug of first choice for most rickettsial illness. It is administered orally and given twice a day. The usual dose is 100 mg PO bid. Minocycline can be used to eradicate the meningococcal carrier state although rifampin is currently preferred. Tigecycline has an extended spectrum that includes methicillin-resistant staphylococci and vancomycin-resistant enterococci (VRE). It also has a broad gram-negative spectrum including some organisms that are resistant to other tetracyclines but both *Proteus* and *Pseudomonas* species are resistant. Tigecycline is approved for treatment of skin and soft tissue infections and is often used to treat MRSA or VRE. It does not

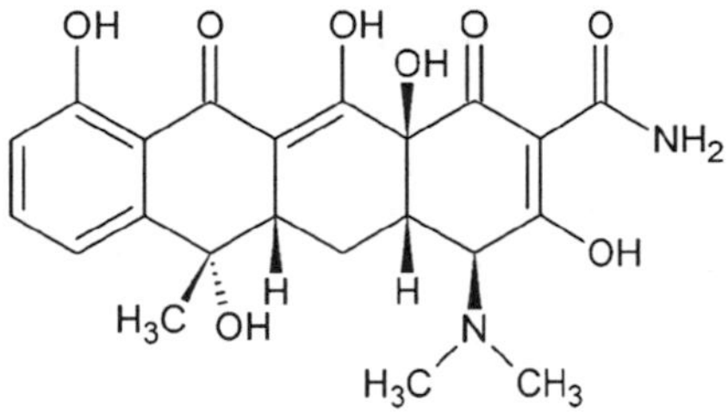

Fig. 13.3 Structure of doxycycline

concentrate well in the urine, and therefore is not an optimal drug for urinary tract infections. The use of tigecycline to treat pneumonia in immunocompromised patients has been disappointing (Fig. 13.3).

Common Uses

First-line drug for rickettsial diseases
Mycoplasma pneumoniae, *Chlamydia*
Useful for treating *Vibrio* (if sensitive)
In combination with other drugs for *H. pylori*
Second choice agent for Lyme disease or syphilis
Tigecycline can be used for treatment of resistant gram-positive and gram-negative organisms (but not *Pseudomonas*)

Adverse Effects

The most common reasons for discontinuing tetracyclines are nausea, vomiting, and/or diarrhea. These side effects can often be controlled by administration with food or decreasing the dosage. Administration of tetracyclines is contraindicated in children less than 8 years old and in pregnant women because it binds calcium and can lead to fluorescence and discoloration of teeth, as well as enamel and bone complications. These drugs have been associated with liver toxicity, particularly in pregnant women. Photosensitization is associated with administration of tetracyclines, and vestibular reactions including dizziness and vertigo have been seen, particularly with high doses of doxycycline. High doses of doxycycline (200 mg) given after a tick bite to prevent Lyme disease are associated with gastrointestinal side effects in one-third of people being treated. It is usually very well tolerated at conventional doses (100 mg bid).

Further Reading

Chemaly, R. F., Hanmod, S. S., Jiang, Y., Rathod, D. B., Mulanovich, V., Adachi, J. A., et al. (2009). Tigecycline use in cancer patients with serious infections: a report on 110 cases from a single institution. Medicine (Baltimore), 88(4), 211–220.

Edson, R. S., Bundrick, J. B., & Litin, S. C. (2011). Clinical pearls in infectious diseases. Mayo Clin Proc, 86(3), 245–248.

Roberts, J. A., & Lipman, J. (2009). Pharmacokinetic issues for antibiotics in the critically ill patient. Crit Care Med, 37(3), 840–851; quiz 859.

Smilack, J. D. (1999). The tetracyclines. Mayo Clin Proc, 74(7), 727–729.

Tigecycline (tygacil). (2005). Med Lett Drugs Ther, 47(1217), 73–74.

Part II
The Clinician Approaches a Patient with an Infectious Disease

Chapter 14
Clinical Approach to the Treatment of Infectious Diseases

The Four Questions

A clinician approaching a patient with an infectious disease needs to consider both the host and the pathogen. Different hosts may be more or less likely to be infected with a particular pathogen. Therefore, the clinician needs to pay considerable attention to the status of the host. Is the patient someone who has just received chemotherapy and therefore has no circulating neutrophils? If so, this patient will be predisposed to aerobic gram-negative and gram-positive bacterial infections and should be treated even if minimal signs of infection are present. On the other hand, if this patient is a normal host, one would expect the infected individual to manifest signs and symptoms characteristic of a localized infection. For example, most people presenting with bacterial pneumonia will have the classic triad of fever, productive cough, and signs of consolidation on physical examination and chest X-ray. Most of these people will have pneumonia caused by *S. pneumoniae*, making the therapeutic decisions about what antibiotic to use very easy. In the case of a patient without circulating neutrophils, productive cough or sputum will be absent because in order to generate sputum or have a productive cough, one needs neutrophils. Similarly, neutrophils that migrate to the site of the bacterial infection initiate the release of inflammatory cytokines that cause the local accumulation of fluid that leads to changes on chest X-rays and related findings on physical examination. The clinician will hear the consequences of fluid in the lung as "signs of consolidation," usually detected as egophony, on physical exam.

Similarly, a patient without T cell immunity (as a result of chemotherapy, an acquired or congenital abnormality, or infection with HIV-1) will have an atypical response to intracellular organisms such as *M. tuberculosis* and will fail to wall off the organism. These situations make clinical diagnoses of infections much more difficult and require a different approach on the part of the clinician.

In the case of the neutropenic patient, it is important for the clinician to treat the patient with a fever even if there are no localizing signs of infection. Patients with severe neutropenia are likely to be infected with organisms that normally inhabit the

R.W. Finberg and R. Guharoy, *Clinical Use of Anti-infective Agents: A Guide on How to Prescribe Drugs Used to Treat Infections*,
DOI 10.1007/978-1-4614-1068-3_14, © Springer Science+Business Media, LLC 2012

skin or gut, so the approach should be to treat for a broad spectrum of gram-positive and gram-negative aerobic organisms and to continue treatment for the duration of the neutropenic period. Patients who are anticipated to have long periods of neutropenia as well as patients who are deficient in T cell immunity are at risk for certain infections so prevalent that anti-infective agents are given prophylactically. Severely neutropenic patients are routinely given broad-spectrum agents like fluoroquinolones to prevent serious infections from developing. Patients with impaired T cell immunity are routinely treated with trimethoprim/sulfamethoxazole to prevent pneumonia from *P. jirovecii*.

Asking the Four Questions

In approaching a patient with an infection the clinician should always ask four critical questions: "Who?," "Where?," "When?," and "What is in the community?" These four questions are essential to address prior to making any recommendations about anti-infective therapy. However, it is always worth remembering that the single most important question to ask when considering an anti-infective agent of any kind is: "What are we treating (preventing)?" The following outlines the rationale for the other questions:

1. *"Who?"* The first question concerns the host. To the potentially invading microbe, not all people are created equal. People who have no T cells are predisposed to develop infections with intracellular organisms (including intracellular bacteria like *M. tuberculosis* and *Listeria*). People who have antibody deficiencies will be uniquely susceptible to encapsulated bacteria (such as *S. pneumoniae*, *H. influenzae*, and *N. meningitidis*). People with low neutrophil counts are predisposed to develop infections with aerobic bacteria colonizing the skin or GI tract. Predisposition to certain infections also occurs in patients with abnormalities (and sometimes even allelic differences) in certain genes important in innate immunity.
2. *"Where?"* Certain diseases only occur in defined geographic areas. There is no point in treating someone who has never left the East coast of the US for blastomycosis, a disease caused by a fungus that is only seen in the Midwest portion of the US and only found in people who have been to an area of the country where the disease is endemic.
3. *"When?"* Diseases vary by the season. In the Northern Hemisphere, respiratory viruses like respiratory syncytial virus and influenza are common in the winter, and very uncommon in the summer. *Legionella* (a bacterium often associated with standing water) is much more common in the summertime. Tick borne diseases, such as Lyme disease, are spread when ticks are abundant (in the spring and summer).
4. *"What is epidemic in the community?"* A person who presents with fever, headache, and a sore throat in the winter in the midst of an influenza epidemic most

likely has influenza. On the other hand, a person presenting with the same symptoms in the summer in the midst of an outbreak of Coxsackie A virus epidemic, likely has Hand, Foot, and Mouth Disease.

The Remainder of the History

Prior to prescribing an anti-infective agent, the clinician needs to determine if the patient has ever been treated with a given anti-infective agent and find out if there was any history of allergy or other untoward reaction to the agent. Determining whether the report of an allergy represents a serious immunological event or a rash that might have occurred for some other reason (a viral illness or in response to another medication) can be extremely difficult (see Chap. 15 for a discussion on the approach to allergies).

Other drug histories are important to consider. These include prior chemotherapy for cancer patients or the use of anticytokine drugs (such as the TNF inhibitors) that may impair host defenses and make the patient susceptible to infections with intracellular organisms. Corticosteroids may inhibit cellular responses as well, and anti-B cell antibodies such as rituximab prevent the host from making an antibody response to infections. Drugs that stimulate the P450 system may affect metabolism of anti-infective agents, and vice versa.

Establishing a history of renal or liver failure is important given that many anti-infective agents are metabolized by either the liver or kidney.

Specific Issues in the Physical Examination

A physical examination is important in the diagnosis and treatment of patients with infections. Clues to a diagnosis of infection include careful attention to skin lesions, petechiae, and the pattern of distribution of rashes. In addition, the clinician should take note of the patient's height and weight as these are important considerations for predicting drug distribution to tissues and are critical for determining how to dose an anti-infective effectively and without toxicity.

Use of the Laboratory

The ultimate determination as to whether a patient has an infection, and the identification of the pathogen, routinely requires laboratory confirmation by culture, PCR, or antigen testing. In some circumstances, histopathology may be required to establish an infection. Interpretation of laboratory results may be essential to placing the patient in a group that qualifies for treatment or prophylaxis with antimicrobials. For example, patients with HIV-1 whose levels of CD4 T cells are less than 200 are routinely given prophylactic trimethoprim/sulfamethoxazole to prevent pneumonias caused by *P. jirovecii*. Bone marrow transplant recipients are routinely screened for the presence of cytomegalovirus (CMV) by testing for CMV DNA in the serum by PCR. If the virus is identified in the serum at high titer, therapy is initiated. In these cases the laboratory value, rather than the history, leads to the decision to start a specific treatment with an anti-infective agent.

Anyone who prescribes drugs needs to be seriously aware of any condition that would affect the excretion or metabolism of the agent administered. It is prudent to check creatinine clearance as an indication of renal excretion ability, as well as liver function tests since many drugs are metabolized in the liver. Better predictions about drug metabolism will be possible in the future with the increased availability of testing for genetic polymorphisms critical for metabolism.

Chapter 15
Understanding Drug Allergies and Drug Toxicities

Antibiotic Allergy

Clinicians are often confronted with choosing an antibiotic to treat an infection in a patient whose medical record carries an allergy warning. On one hand, allergic reactions are associated with significant morbidity and mortality and increased health care costs. On the other hand, using substitute antibiotics for patients without evidence of true allergies contributes to inappropriate use of broad-spectrum agents, increased costs, and resistance. Discrepancies between medical records and reviews based on patient interviews have questioned the validity of allergy information obtained from medical records. The term antibiotic "allergy" is often inappropriately used to characterize any adverse drug reaction.

Classification of Antibiotic Allergy

By definition, an allergic reaction involves an immunologic reaction to an agent. IgE-mediated reactions represent only a small subgroup. These reactions include any combination of sudden hypotension, bronchospasm, hyperperistalsis, erythema, pruritus, urticaria, angioedema, and/or arrhythmias. The reactions usually start within 15 min after antibiotic administration. Most IgE-mediated reactions do not result in anaphylaxis.

Most reactions are non-IgE-mediated. Serum sickness-like reactions are rare and can occur via immune complex mediation. Thrombocytopenia, hemolysis, interstitial nephritis, and neutropenia can result from antibiotic mediation. Nonurticarial rashes may result from delayed hypersensitivity reactions. T cells play a major role in delayed hypersensitivity reactions, including antibiotic-induced maculopapular eruptions. Immunopathologic mechanisms of fixed drug eruptions, autoimmune disease, erythema multiforme, antibiotic fever, rash, and lymphadenopathy are not clear.

R.W. Finberg and R. Guharoy, *Clinical Use of Anti-infective Agents: A Guide on How to Prescribe Drugs Used to Treat Infections*,
DOI 10.1007/978-1-4614-1068-3_15,

Clinical Features

The most common reactions include skin eruptions, urticaria, pruritus, and maculopapular skin eruptions and occur between days to weeks after initial exposure to the agent. On some occasions, the reactions occur much sooner after secondary exposure within minutes to hours. Hypersensitivity reactions include eosinophilia, fever, and other cutaneous reactions.

Diagnosis of Antibiotic Allergy

First and foremost, a close review of patient history is the best and only tool for making the diagnosis. A thorough interview with the patient can help to ascertain whether the patient had IgE-mediated reactions, such as pruritus, angioedema, urticaria, hypotension, bronchospasm, hyperperistalsis, and/or arrhythmia in the past.

Diagnostic tests are limited. Skin testing can be used to detect specific IgE antibodies. However, the tests are not standardized with the exception of penicillin. The majority of the literature describes penicillin allergy since many patients claim to have allergy to penicillin. There are no valid reagents to test specific IgE antibodies. Use of a specific agent may be used for allergic testing. However, a negative reaction may not validate absence of IgE antibodies.

Skin testing for penicillin allergy is highly accurate and involves both an epidermal and an intracutaneous prick test. The products of penicillin metabolism are termed major and minor determinants and represent the amount of drug metabolized. Skin testing for major and minor determinants identifies 97–99% of potential reactors. The antigenic determinants include both the major determinant, benzylpenicilloyl polylysine, and multiple minor determinants (Fig. 15.1). The goal is to inject all metabolites with the potential to cause an allergic reaction. Testing is performed with penicilloyl polylysine and either penicillin diluted to 10,000 units per milliliter or a mixture of minor determinants which includes a mixture of benzyl-*n*-propylamine, benzyl penicilloate, and benzyl penilloate. Skin prick test is done first and intracutaneous testing is followed if there are no reactions within 15 min. Erythema and a 3 mm increase in wheal diameter (compared with a negative control) are considered a positive test. A negative test confirms that the previous test was not IgE-mediated or the antibodies are absent offering the option of penicillin administration to a patient. Patients with a negative skin test and past history of penicillin allergy have on average less than 1% chance of developing an IgE-mediated reaction. However, skin testing should not be done immediately after anaphylaxis occurs because of the potential of a period of anergy after mast cell degranulation. Penicilloyl polylysine is not currently commercially available in the US because of high manufacturing costs for a very limited market. Production is expected to resume in the near future. Fluorescent ELISA can be utilized to measure penicillin specific IgE levels as an alternative to skin testing. However, it has limited value since it can only determine antibodies to the major determinants.

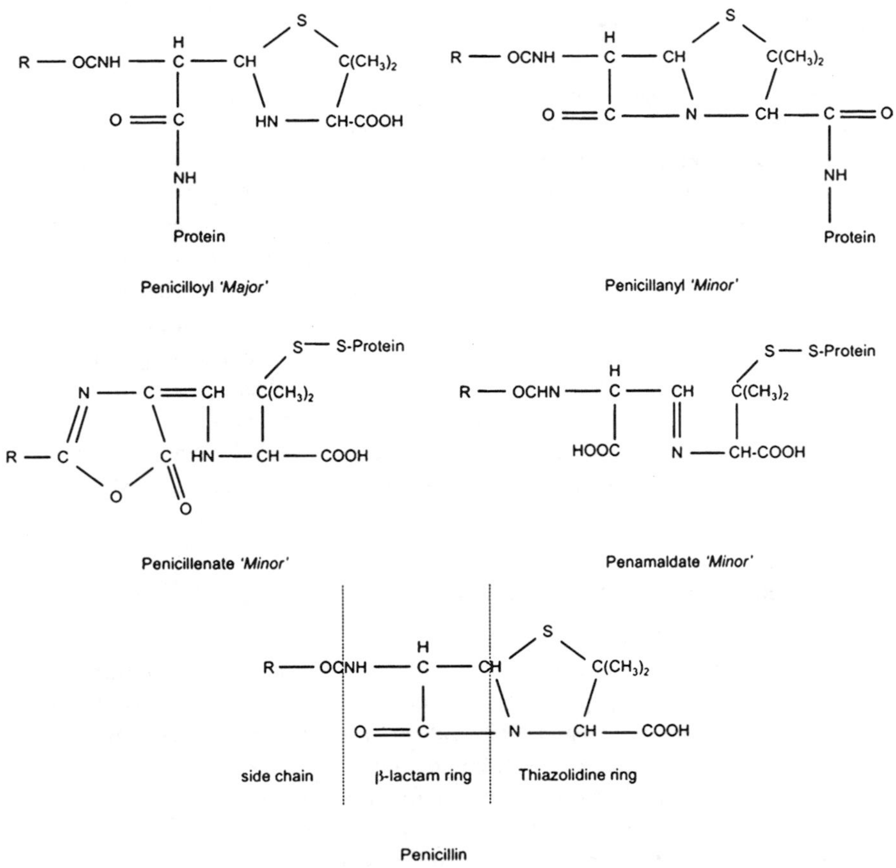

Penicilloyl *'Major'*

Penicillanyl *'Minor'*

Penicillenate *'Minor'*

Penamaldate *'Minor'*

side chain | β-lactam ring | Thiazolidine ring

Penicillin

Fig. 15.1 Penicillin major and minor determinants (Adapted with permission from Baldo, B.A., Pham, N.H. (1994). Structure–activity studies on drug-induced anaphylactic reactions. *Chem Res Toxicol.* Nov-Dec; 7(6):703-21. Copyright (1994) American Chemical Society

Other Methods

Antibiotic patch testing that requires observation of local reactions after topical application of an antibiotic on a pad under hypoallergenic occlusive tape for 48–72 h is not accepted as a standard of practice in North America. It may be useful in predicting non-IgE mediated reactions. More research in this area is needed.

A positive Coombs' test indicates the presence of cell-bound antibodies which results in penicillin-induced hemolytic anemia. Drug-specific T tests may be detected with lymphocyte transformation tests. A positive test indicates that an agent is sensitized to the drug. However, this test is not available in the US.

Differential Diagnosis

Many patients experience drug-related adverse effects rather than a true allergic reaction. One must critically examine the patient's drug regimen and conduct a thorough investigation. Some infections may activate T cells in the presence of a specific agent which may lead to hypersensitivity reactions. A rash that occurs after administration of amoxicillin to a patient with Epstein–Barr virus infection is not a manifestation of an amoxicillin allergy in the patient. Epstein–Barr virus is known to activate T cells that lead to post-amoxicillin rashes. This is not a true allergy since the same patient, when challenged again with amoxicillin, will not develop a rash. The "red man syndrome" associated with vancomycin use looks like a rash caused by an allergic response, but it is caused by histamine release and the reaction is related to rapid infusion resulting in direct release of mediators from mast cells. Administration of the same drug to the same patient over a longer period of time solves the problem, as histamine is not released when the drug is given slowly.

Cross-Reactivity Between Cephalosporins and Penicillins

Penicillin and cephalosporins share the same beta-lactam structure and the potential of cross-reactivity exists. It appears that the cross-reactivity depends on the similarity of side chain structures. Cephalothin and cephaloridine have greater cross-sensitivity with penicillin because of a side chain similar to that of benzylpenicillin. The same phenomenon is not seen with cefuroxime and cefotaxime because of differences in side chains. Studies have reported that cross-reactivity between first- and second-generation cephalosporins and penicillin with similar side chains has an incidence between 14 and 38%. Based on published literature, cross-reactivity between penicillin and cephalosporins with different side chains is <10%. Initial reports of reactions between penicillins and cephalosporins were overestimated as a result of the use of contaminated preparations used for testing. Although there is cross-reactivity between first-generation cephalosporins (e.g., cefazolin) and penicillins, there is minimal cross-reactivity between penicillins and third-generation cephalosporins and recent studies indicate that third-generation cephalosporins can be used in patients with a history of penicillin allergy (see Pichichero 2005).

Cross-Reactivity Among Cephalosporins

Cephalosporins are an important therapy for many infectious diseases and hence the use of an agent is challenging when a penicillin or another cephalosporin has been implicated in causing an allergic reaction. Cephalosporins have two side chains (Fig. 15.2) in C7 in R1 and different substitutions in C3 position (R2). Similarities

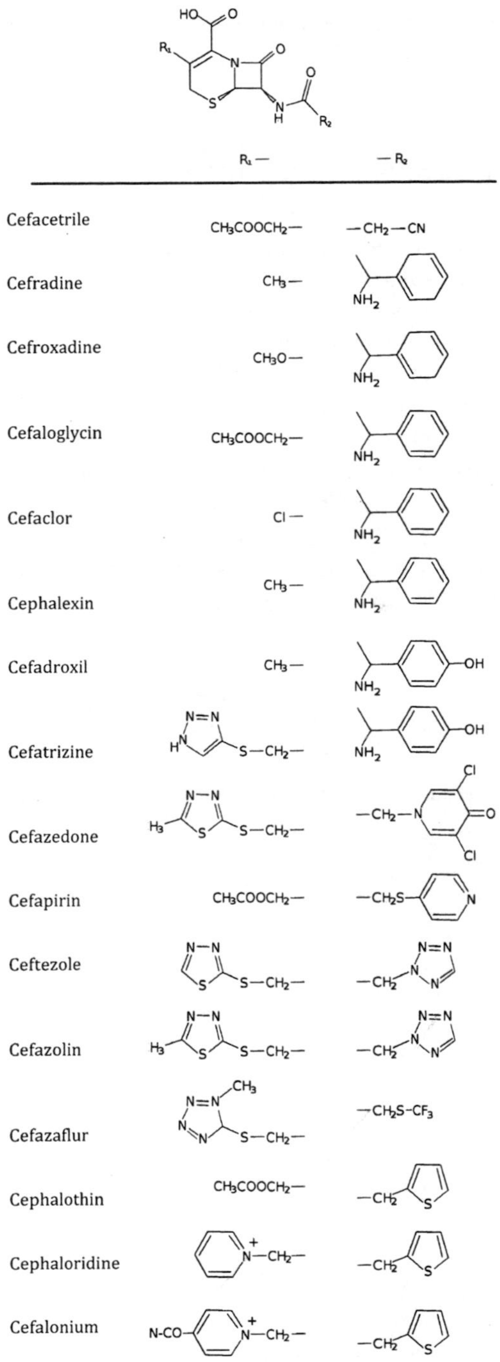

Fig. 15.2 Structures of cephalosporins

	R_1—	—R_2
Cefacetrile	CH_3COOCH_2—	—CH_2—CN
Cefradine	CH_3—	NH_2
Cefroxadine	CH_3O—	NH_2
Cefaloglycin	CH_3COOCH_2—	NH_2
Cefaclor	Cl—	NH_2
Cephalexin	CH_3—	NH_2
Cefadroxil	CH_3—	NH_2 —OH
Cefatrizine	N=N, H N, S—CH_2—	NH_2 —OH
Cefazedone	N—N, H_3, S, S—CH_2—	—CH_2—N, Cl, =O, Cl
Cefapirin	CH_3COOCH_2—	—CH_2S—, N
Ceftezole	N—N, S, S—CH_2—	N=N, —CH_2—N, N
Cefazolin	N—N, H_3, S, S—CH_2—	N=N, —CH_2—N, N
Cefazaflur	CH_3, N=N, N, N, S—CH_2—	—CH_2S-CF_3
Cephalothin	CH_3COOCH_2—	—CH_2, S
Cephaloridine	N^+—CH_2—	—CH_2, S
Cefalonium	N-CO, N^+—CH_2—	—CH_2, S

in any of them may lead to cross-reactivity among the agents. In vivo cross-reactivity studies are limited.

Penicillin Allergy and Carbapenems

A five-membered ring attached to the beta-lactam ring in the carbapenem structure is presumed to be responsible for cross-reactivity between the classes. Cross-sensitivity between penicillin and carbapenems has been reported between 0.9 and 47% in the literature. Variability in study designs is believed to have contributed to this wide range. A broad definition for allergic reaction or lack of verification of penicillin allergies was used in some of these studies. The studies which verified penicillin allergy by skin test to major and minor penicillin determinants and tested for carbapenem allergy by administering a full therapeutic dose to carbapenem skin test-negative patients reported cross-reactivity of around 1% (see Romano et al. 2006, 2007; Atanaskovic-Markovic et al. 2008). Based on these studies, carbapenem may be administered to patients with penicillin allergy who are negative for skin tests to carbapenem (Fig. 15.3).

Penicillin and Monobactams

The monobactam structure is composed of a beta-lactam structure without adjoining ring or side chain (Fig. 15.4).

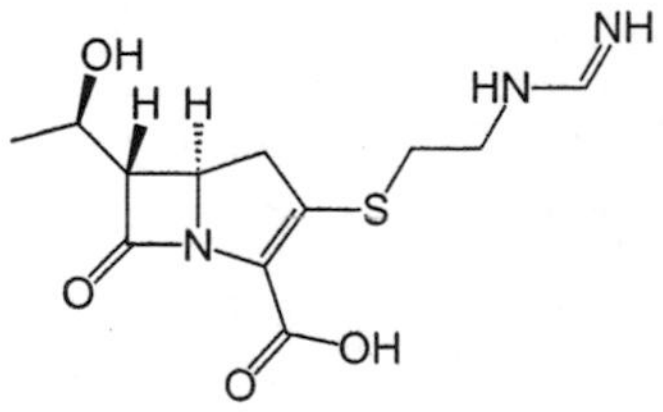

OH H H NH HN S N O OH O

Fig. 15.3 Structure of imipenem (a carbapenem)

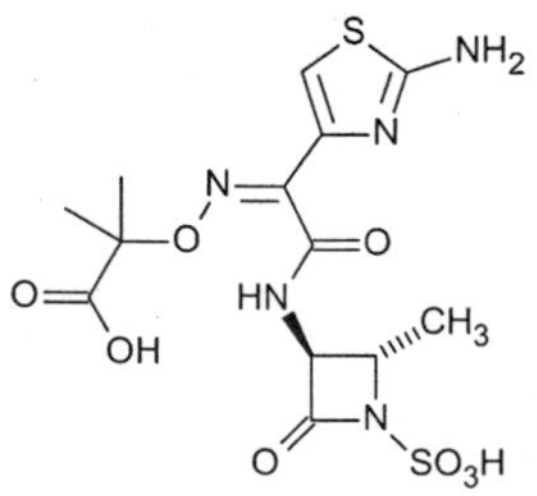

S NH_2 N N O O HN CH_3 O OH N O SO_3H

Fig. 15.4 Monobactam structure

Because of the unique structure, monobactams can be used in patients with penicillin allergy. Aztreonam is the only commercially available product in this class. Studies have shown that anti-aztreonam antibodies' cross-reactivity with penicillins and cephalosporins is negligible. Studies have also suggested that aztreonam antibodies were directed to the side chain rather than the beta-lactam ring. Aztreonam is weakly immunogenic with minimal cross-reaction with penicillin and cephalosporin antibodies. A case report addressed the concern of cross-reactivity between ceftazidime and aztreonam because of the structural similarities between them. In this case, the patient was allergic to ceftazidime and aztreonam, not benzylpenicillin, amoxicillin, or other cephalosporins (see Perez Pimiento et al. 1998). Additionally, studies have reported 16.7% of cystic fibrosis patients allergic to ceftazidime became sensitized to aztreonam and developed angioedema and bronchospasm upon reexposure to aztreonam (see Moss 1991; Moss et al. 1991). Aztreonam should be administered with caution in patients with hypersensitive reactions to ceftazidime.

Sulfonamide Allergy

Clinicians are often challenged with choice of therapy in patients with "sulfa" allergy because of potential cross-reactivity between sulfonylarylamines and nonsulfonylarylamines and sulfonamide moiety-containing drugs. There are three distinct groups which belong to the sulfonamides (see Figs. 15.5 and 15.6).

Nonsulfonylarylamines

The first group has a sulfonamide moiety connected to a benzene ring or other cyclic structure without the amine moiety at the N-4 position. These are the nonsulfonylarylamines. Medications belonging to this group include carbonic anhydrase inhibitors, sulfonylureas, loop diuretics, thiazides, cox-2 inhibitors, and protease inhibitors (see Fig. 15.5).

Drugs in Which the Sulfonamide Moiety Is Not Connected to the Benzene Ring

The sulfonamide moiety is not directly connected to the benzene ring in the second group whereas it is directly connected to a benzene ring with an unsubstituted amine ($-NH_2$) at the N4 position (sulfonylarylamines) in the third group. The drugs included in the second group include 5-hyroxytryptamine-1 (5-HT-1) agonists, topiramate, zonisamide, sotalol, probenecid, and ibutilide.

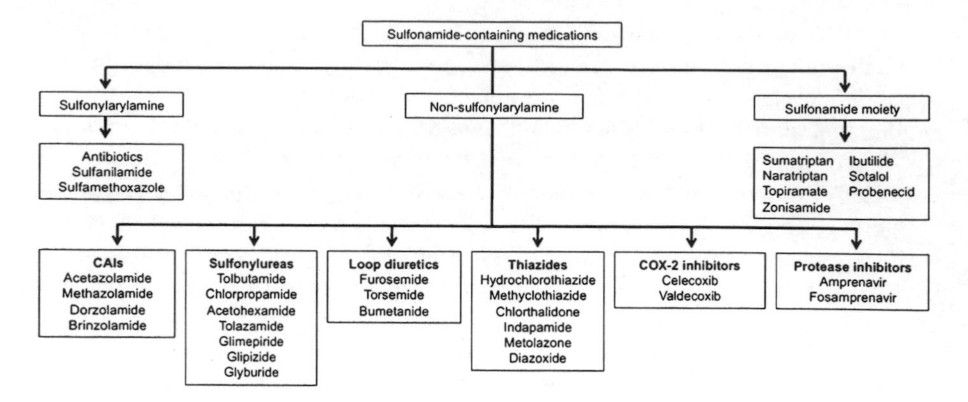

Fig. 15.5 Sulfonamide containing drugs. *CAI* carbonic anhydrase inhibitors; *cox-2* cyclooxygenase-2; *thiazides* thiazide diuretics and related drugs (adapted with permission from the *Annals of Pharmacotherapy*; 2005; 39:290–301)

Sulfonylarylamine Drugs

The third group, sulfonylarylamines, includes the antibiotics sulfanilamide and sulfamethoxazole.

In a recently published analysis of the literature by Johnson et al., the authors commented that the cross-reactivity between sulfonylarylamine drugs and other sulfa drugs is not supported by data since many of the reported case reports do not conclusively support either a connection or an association between the observed cause and effect. The authors recommended that nonsulfonylarylamine or sulfonamide moiety-containing drugs can be administered to patients with a nonsevere sulfonylarylamine allergy with appropriate monitoring.

Multiple Antibiotic Allergy

It is extremely difficult for a clinician to determine therapeutic options in the event of allergies to multiple classes of agents. Clinicians should note that serious allergies are rare with non-beta-lactam agents. Antihistamines can be used for nonurticarial rashes. Desensitization may be considered in the event of a patient requiring an agent with previous history of IgE-mediated reaction.

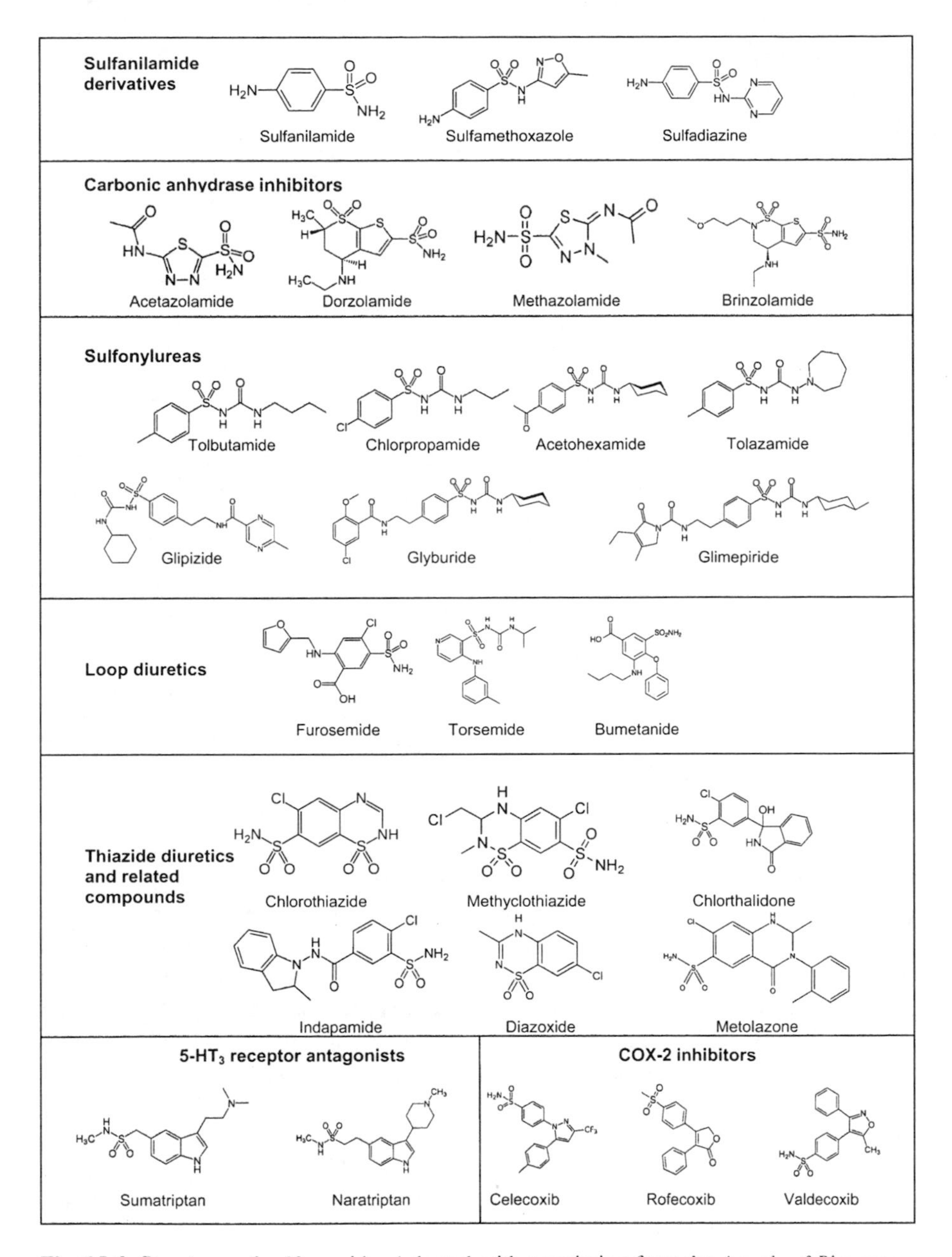

Fig. 15.6 Structures of sulfonamides (adapted with permission from the *Annals of Pharmacotherapy*; 2005; 39:290–301)

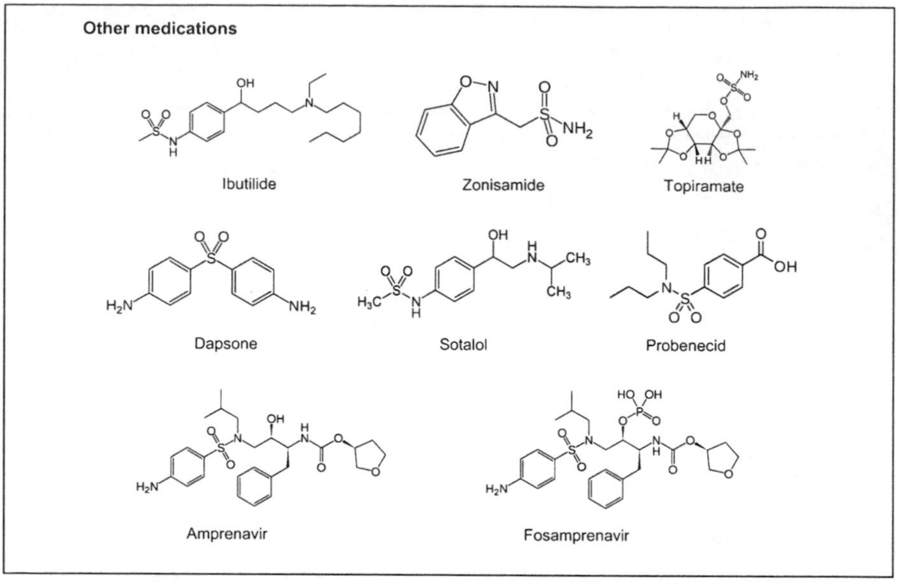

Other medications

Ibutilide Zonisamide Topiramate

Dapsone Sotalol Probenecid

Amprenavir Fosamprenavir

Fig. 15.6 (continued)

Treatment Options

A careful assessment and thorough patient history are the most crucial tools to identify exact nature of the reaction that resulted in the immune reaction. Information on previous tolerance to suspected agent or other beta-lactam agent is critical since it provides clues to which agent sensitized the patient. The patient history should include a history of all medications including over-the-counter and herbal products the patient was taking at the time of the adverse event. The American Academy of Allergy, Asthma and Immunology has developed practice guidelines based on evidence and expert opinion (see The diagnosis and management of anaphylaxis: an updated practice parameter 2005; Executive summary of disease management of drug hypersensitivity: a practice parameter. Joint Task Force on Practice Parameters, the American Academy of Allergy, Asthma and Immunology, the American Academy of Allergy, Asthma and Immunology, and the Joint Council of Allergy, Asthma and Immunology 1999).

Desensitization

Drug desensitization is an option for patients with IgE-mediated reactions who require the offending agent. Expert personnel in the hospital setting need to conduct desensitization, which includes administration of increasing amounts of antibiotic slowly over a period of hours until a therapeutic dose is reached. The administration starts with a very low dose (micrograms) and doses are doubled every 15–30 min. Patients are closely monitored and the procedure is discontinued in the event of a severe adverse reaction such as hypotension or severe bronchospasm.

Rechallenge Test

This method can be used for patients that experienced non-IgE-mediated reactions. Initial doses are significantly higher than desensitization test and the interval between doses varies from hours to days. Administration is discontinued in the event of reactions such as hemolytic anemia, toxic epidermal necrolysis, immune complex reactions, and Stevens–Johnson syndrome.

Conclusion

A critical review of patient history can provide the clue to whether the event was immunologic or not. Skin testing, if available can be used in patients with IgE-mediated reactions. A beta-lactam agent can be used in the event that the skin testing is negative and they should be avoided in patients with a positive result. A desensitization procedure may be considered for the patients with positive skin testing who absolutely require the specific agent.

Further Reading

Adkinson, N. F., & Middleton, E. (2003). Middleton's allergy : principles & practice (6th ed.). Philadelphia, Pa.: Mosby.

Atanaskovic-Markovic, M., Gaeta, F., Medjo, B., Viola, M., Nestorovic, B., & Romano, A. (2008). Tolerability of meropenem in children with IgE-mediated hypersensitivity to penicillins. Allergy, 63(2), 237–240.

The diagnosis and management of anaphylaxis: an updated practice parameter. (2005). J Allergy Clin Immunol, 115(3 Suppl 2), S483-523.

Executive summary of disease management of drug hypersensitivity: a practice parameter. Joint Task Force on Practice Parameters, the American Academy of Allergy, Asthma and Immunology, the American Academy of Allergy, Asthma and Immunology, and the Joint Council of Allergy, Asthma and Immunology. (1999). Ann Allergy Asthma Immunol, 83(6 Pt 3), 665–700.

Gruchalla, R. S., & Pirmohamed, M. (2006). Clinical practice. Antibiotic allergy. N Engl J Med, 354(6), 601–609.

Johnson, K. K., Green, D. L., Rife, J. P., & Limon, L. (2005). Sulfonamide cross-reactivity: fact or fiction? Ann Pharmacother, 39(2), 290–301.

Lee, C. E., Zembower, T. R., Fotis, M. A., Postelnick, M. J., Greenberger, P. A., Peterson, L. R., et al. (2000). The incidence of antimicrobial allergies in hospitalized patients: implications regarding prescribing patterns and emerging bacterial resistance. Arch Intern Med, 160(18), 2819–2822.

Macy, E. (2004). Multiple antibiotic allergy syndrome. Immunol Allergy Clin North Am, 24(3), 533–543, viii.

Moss, R. B. (1991). Sensitization to aztreonam and cross-reactivity with other beta-lactam antibiotics in high-risk patients with cystic fibrosis. J Allergy Clin Immunol, 87(1 Pt 1), 78–88.

Moss, R. B., McClelland, E., Williams, R. R., Hilman, B. C., Rubio, T., & Adkinson, N. F. (1991). Evaluation of the immunologic cross-reactivity of aztreonam in patients with cystic fibrosis who are allergic to penicillin and/or cephalosporin antibiotics. Rev Infect Dis, 13 Suppl 7, S598-607.

Perez Pimiento, A., Gomez Martinez, M., Minguez Mena, A., Trampal Gonzalez, A., de Paz Arranz, S., & Rodriguez Mosquera, M. (1998). Aztreonam and ceftazidime: evidence of in vivo cross allergenicity. Allergy, 53(6), 624–625.

Pichichero, M. E. (2005). A review of evidence supporting the American Academy of Pediatrics recommendation for prescribing cephalosporin antibiotics for penicillin-allergic patients. Pediatrics, 115(4), 1048–1057.

Robinson, J. L., Hameed, T., & Carr, S. (2002). Practical aspects of choosing an antibiotic for patients with a reported allergy to an antibiotic. Clin Infect Dis, 35(1), 26–31.

Romano, A., Viola, M., Gueant-Rodriguez, R. M., Gaeta, F., Pettinato, R., & Gueant, J. L. (2006). Imipenem in patients with immediate hypersensitivity to penicillins. N Engl J Med, 354(26), 2835–2837.

Romano, A., Viola, M., Gueant-Rodriguez, R. M., Gaeta, F., Valluzzi, R., & Gueant, J. L. (2007). Brief communication: tolerability of meropenem in patients with IgE-mediated hypersensitivity to penicillins. Ann Intern Med, 146(4), 266–269.

Chapter 16
Principles of Antibiotic Resistance

Because of their short dividing time (*E. coli* replicate approximately every 30 min) and their ability to accept DNA from other bacteria or phages, most bacteria are capable of becoming resistant to antibiotics very quickly. The history of antibacterial therapy is characterized by the fact that every time a new antibacterial agent is introduced, the organisms become resistant. The exception to this rule includes syphilis which has remained sensitive to penicillin for over 50 years since the introduction of this agent. *N. gonorrhea* and *S. aureus*, which became resistant within a few years after the introduction of penicillin, represent the more common response of bacteria to the introduction of an agent.

The mechanisms used by bacteria to develop antimicrobial resistance include alteration of ribosomes (in the case of ribosomally targeted drugs), alteration of cell wall-binding proteins (in the case of beta-lactam antibiotics that target the cell wall), expression of enzymes that degrade antibiotics (in the case of aminoglycoside resistance and beta-lactam resistance that is mediated by beta-lactamases), and alterations of the cell wall that prevent entry or promote excretion of the drug (antibiotic "pumps") (for a detailed drug by drug list of mechanisms see Shlaes et al. 1997).

Resistance can emerge in one of two ways (1) as a result of selection by the antimicrobial agent for a bacterium that is resistant or (2) as a result from genetic transfer from another bacterium or a phage. The consequence of the second method of transfer is that resistance can be rapidly spread and resistant organisms may be found in patients not previously exposed to a given antibiotic. The rapid spread of resistance in gram-negative organisms was facilitated by plasmids that carried resistance genes and spread from one bacterium to another by conjugation.

In vitro, the ideal condition for bacteria to develop resistance to an antibiotic is in culture with suboptimal doses of drug. This will select for variants that are more resistant to the drug. Continued antibiotic pressure also leads to selection of resistant variants in vivo. Thus, when using antibiotics, one should always use adequate doses and treat for the shortest period of time necessary to clear the infection. Unlike drugs that target mammalian receptor proteins, most antimicrobials are safe to give in large doses since they do not damage host cells. Therefore, it is not usually wise

R.W. Finberg and R. Guharoy, *Clinical Use of Anti-infective Agents: A Guide on How to Prescribe Drugs Used to Treat Infections*,
DOI 10.1007/978-1-4614-1068-3_16,

to use minimal doses of antimicrobial agents. Use of antibiotics for prolonged periods is also not wise because it leads to selection of both resistant organisms and recolonization with unusual organisms (such as *C. difficile*) that may have negative consequences for the host.

Common problems in antibiotic prescribing practices include the use of antibiotics with a spectrum that is broader than what is necessary and the treatment of colonizing microbes that are not causing true infections. An example is the treatment of asymptomatic bacteriuria. Treatment of bacteria in the urine in the absence of signs or symptoms is only indicated before genitourinary procedures and in pregnant women or immunocompromised hosts.

The following approaches to drug prescribing are recommended to minimize the development of resistant bacteria:

1. Never use antimicrobials without a clinical indication.
2. Treat for the shortest possible time necessary to eliminate the organism.
3. Always treat with a dose of drug that is adequate to kill all the bacteria (do not under-dose).
4. Empiric and prophylactic use of antimicrobials should be carefully monitored and only used when there is clear evidence for their efficacy.

Further Reading

Dellit, T. H., Owens, R. C., McGowan, J. E., Jr., Gerding, D. N., Weinstein, R. A., Burke, J. P., et al. (2007). Infectious Diseases Society of America and the Society for Healthcare Epidemiology of America guidelines for developing an institutional program to enhance antimicrobial stewardship. Clin Infect Dis, 44(2), 159–177.

Shlaes, D. M., Gerding, D. N., John, J. F., Jr., Craig, W. A., Bornstein, D. L., Duncan, R. A., et al. (1997). Society for Healthcare Epidemiology of America and Infectious Diseases Society of America Joint Committee on the Prevention of Antimicrobial Resistance: guidelines for the prevention of antimicrobial resistance in hospitals. Clin Infect Dis, 25(3), 584–599.

Chapter 17
Clinical Approach to Treatment of Bacterial Infections

Approaching the Patient with a Bacterial Infection

The determination that a patient has a bacterial infection can be made on the basis of the history, physical examination, and clinical setting, or it can be made on the basis of either histopathologic or microbiologic diagnostic tests. Histologic analysis of infected tissues will often yield information as to the staining characteristics of the bacteria. Even before the microbiology laboratory identifies the strain of bacteria, the organism will be reported as gram negative or gram positive. Clinical clues can also suggest whether the likely infection is caused by gram-negative or gram-positive organisms. For example, infections of the skin are likely to be caused by *Streptococcus* or *Staphylococcus*, both of which are aerobic gram-positive organisms. Infections of the gastrointestinal tract, on the other hand, are likely to be caused by either aerobic or anaerobic gram-negative organisms since these are the dominant flora of the gut.

1. *Approach to treatment of infections with gram-positive organisms:*

   (a) *Gram-positive cocci*: While penicillin G was the initial treatment of choice for gram-positive infections, widespread use of this drug has resulted in resistance to most *Staphylococcus* species. Penicillin does remain the treatment of group A streptococci as well as many species of viridans streptococci (a major cause of endocarditis). Most strains of *S. epidermidis* and *S. aureus* are resistant to penicillin G. While the antistaphylococcal penicillins (nafcillin and oxacillin) are excellent choices for treatment of sensitive strains of staphylococci, the widespread prevalence of MRSA, and the natural resistance of many non-*aureus* staphylococci (such as *S. epidermidis*), makes the use of another agent mandatory until sensitivity testing is completed. Vancomycin is the most commonly used agent in this setting. However, if the organism is methicillin sensitive, switching to nafcillin or oxacillin offers advantages in terms of the rapidly bactericidal activities of these drugs. Although bactericidal, vancomycin does not kill sensitive bacteria as well as the antistaphylococcal penicillins do.

R.W. Finberg and R. Guharoy, *Clinical Use of Anti-infective Agents: A Guide on How to Prescribe Drugs Used to Treat Infections*,
DOI 10.1007/978-1-4614-1068-3_17,

While penicillin G is the traditional treatment for pneumococcal pneumonia, many strains are now more resistant to pen G, and a third-generation cephalosporin such as ceftriaxone is often chosen for treatment of pneumococcal pneumonia.

Enterococci cannot be treated with penicillin alone. Treatment of enterococcal endocarditis requires the use of penicillin with an aminoglycoside or vancomycin plus an aminoglycoside. Enterococcal urinary tract infections can be treated with ampicillin if the organism is sensitive. Vancomycin-resistant enterococci (VRE) may be sensitive to daptomycin, linezolid, or quinupristin/dalfopristin. Some enterococci are sensitive to carbapenem antibiotics (imipenem, doripenem).

(b) *Gram-positive rods*: The approach to treatment of infections with gram-positive rods depends very much on the clinical situation. In immunocompromised patients with meningitis, the isolation of a gram-positive rod from blood or CSF should suggest the possibility of *Listeria monocytogenes* meningitis and this would be an indication for treatment of the patient with ampicillin (or in the case of ampicillin allergic patients, with trimethoprim–sulfamethoxazole). *Listeria monocytogenes* is not sensitive to cephalosporins like ceftriaxone (an agent commonly used for treatment of meningitis). In the case of patients exposed to bioterrorists' actions or in contact with the goatskins from the third world, consideration of infection with *Bacillus anthracis* is an issue. Although in the past, most strains of *B. anthracis* were sensitive to penicillin, because this organism may express an inducible beta-lactamase, current recommendations are to treat with a fluoroquinolone (ciprofloxacin or levofloxacin) or doxycycline with the addition of clindamycin (to stop toxin production).

More commonly, clinicians are faced with the problem of treating *Proprionibacterium acnes* infections or infections with diphtheroids. These organisms can cause subacute infections in a variety of clinical settings including IV catheter infections and they are usually treated with vancomycin or penicillin.

2. *Treatment of gram-negative infections:*

(a) *Gram-negative cocci*: Common gram-negative cocci include *N. meningitidis* and *N. gonorrhea. N. meningitidis*, the causal agent of meningococcal pneumonia, remains sensitive to penicillin G and can be treated with penicillin (although the drug should be given very frequently, q 2H or by continuous infusion to ensure optimal pharmacokinetics). Alternatively, *N. meningitidis* can be treated with ceftriaxone, cefuroxime, or ceftazidime. Most strains of *N. gonorrhea*, the causal agent of gonorrhea, are resistant to penicillin but can be treated with ceftriaxone.

(b) *Gram-negative bacilli*: While ampicillin and cephalexin were traditionally used to treat *E. coli*, particularly in the case of urinary tract infections, resistance to these drugs has made them poor choices for initial therapy (unless the organisms are known to be sensitive to these agents). Fluorinated quinolones such as ciprofloxacin are excellent choices for treatment of a broad spectrum of aerobic gram-negative bacilli. Third-generation cephalosporins (e.g., ceftriaxone) or carbapenems (e.g., imipenem, doripenem) also have an excellent spectrum of activity against enteric gram-negative organisms. Anaerobic gram-negative rods such as *Bacteroides fragilis* are conventionally treated with clindamycin or metronidazole, while non-*fragilis Bacteroides* species can be treated with penicillin.

3. *Treatment of anaerobic infections*:

Traditionally, most anaerobic infections, including gram-positive cocci such as anaerobic streptococci, *Peptostreptococcus* spp., as well as anaerobic gram-positive bacilli such as *Clostridium* species, have been treated with penicillin G.

For treatment of anaerobic gram-negative organisms, penicillin also has excellent activity against non-*fragilis Bacteroides* species and therefore is excellent treatment for most aspiration pneumonias (since these result from infection with mouth flora). Clindamycin and metronidazole are alternative agents for treating these penicillin responsive anaerobic infections. While most penicillins and cephalosporins have activity against these same organisms, as do the carbapenems (e.g., imipenem), the aminoglycosides and monobactams do not have activity against anaerobes, and the activity of the quinolones is variable (with only the later generations having any substantial activity – see Chap. 12).

Traditionally, treatment of anaerobic infections with anaerobic gram-positive bacilli, such as clostridia, has involved penicillins with clindamycin or metronidazole as alternatives. Treatment of infections with *Bacteroides fragilis*, a gram-negative bacillus which is the dominant flora of the large intestine, requires a combination of a penicillin and a beta-lactamase inhibitor (e.g., ampicillin and sulbactam), since these are beta-lactamase producers. Alternatively a carbapenem, or clindamycin or metronidazole can be used for treatment of *B. fragilis* infections.

Further Reading

Kumar, A., Safdar, N., Kethireddy, S., & Chateau, D. (2010). A survival benefit of combination antibiotic therapy for serious infections associated with sepsis and septic shock is contingent only on the risk of death: a meta-analytic/meta-regression study. Crit Care Med, 38(8), 1651–1664.

Matlow, A. G., & Morris, S. K. (2009). Control of antibiotic-resistant bacteria in the office and clinic. CMAJ, 180(10), 1021–1024.

Moellering, R. C., Jr. (2010). NDM-1--a cause for worldwide concern. N Engl J Med, 363(25), 2377–2379.

Roberts, J. A., Kruger, P., Paterson, D. L., & Lipman, J. (2008). Antibiotic resistance–what's dosing got to do with it? Crit Care Med, 36(8), 2433–2440.

Chapter 18
Clinical Approach to Treatment of Mycobacterial Infections

Tuberculosis is the most common bacterial infection in the world with approximately two billion people infected (one-third of the world's population). Treatment of tuberculosis dates back to the discovery that isoniazid (INH), a compound discovered in 1912 and found to be effective in treatment of TB in the 1950s. It is still the most commonly used agent, and with rifampin, a part of all first-line regimens. While INH is still recommended for initial therapy in most cases, resistance to INH has been reported, and therefore initial therapy always involves multiple drugs.

1. *Treatment of Mycobacterium tuberculosis:*

    (a) *Treatment of latent TB*: Patients with a positive skin test for *M. tuberculosis* (or a laboratory test indicating a T cell response to *M. tuberculosis*) with no evidence of disease (including a normal physical exam and negative chest X-ray) can be treated with isoniazid (INH) for 9 months (the usual dose is 300 mg PO daily), although twice weekly treatment and 6 month regimens also have efficacy. If there is reason to believe that the organism is INH resistant or the patient cannot tolerate INH, a regimen consisting of 600 mg of rifampin daily for 4 months can be substituted for INH.

    (b) *Treatment of active TB*: Treatment of pulmonary tuberculosis is based on the fact that the burden of organisms is high and that resistance is likely to occur if single drugs are given. Therefore, treatment is commonly started with a four-drug regimen (most commonly isoniazid (INH), rifampin (RIF), pyrazinamide (PZA), and ethambutol (EMB). Therapy with all four drugs is routinely continued for 2 months and INH and RIF alone can be given for another 4 months if the patient has responded and the organisms are sensitive to both drugs. If the infection is caused by organisms that are resistant to the commonly used agents, second-line drugs, including fluoroquinolones, streptomycin, amikacin, cycloserine, ethionamide, capreomycin, and *p*-aminosalicylic acid can be used, depending on sensitivity testing. Extrapulmonary TB may require a longer course of therapy depending on

R.W. Finberg and R. Guharoy, *Clinical Use of Anti-infective Agents: A Guide on How to Prescribe Drugs Used to Treat Infections*,
DOI 10.1007/978-1-4614-1068-3_18,

the location of disease and the rapidity of the response (with 9–12 months recommended for CNS tuberculosis).

2. *Treatment of Mycobacterium leprae*: Leprosy is commonly treated with dapsone and rifampin. In some regimens, clofazimine is added as a third agent. Low organism burden tuberculoid disease may be treated for 6 months but lepromatous disease (which is characterized by a high organism burden) requires a 12-month course of therapy.
3. *Treatment of Mycobacterium kansasii*: *Mycobacterium kansasii*, an endemic disease is resistant to pyrazinamide, and therefore treatment strategies rely on INH, RIF, and EMB. A conventional approach to treatment is to use all three drugs (INH, RIF, and EMB) and start with ethambutol at 25 mg/kg for the first 2 months and then decrease the dose to 15 mg/kg (because of the risk of optic neuritis at high doses) to complete an 18-month course of treatment.
4. *Treatment of Mycobacterium avium-intracellulare*: *Mycobacterium avium-intracellulare* can present as a pulmonary disease, usually in people with underlying lung disease, or as disseminated infection in immunocompromised hosts. Treatment regimens usually include clarithromycin or azithromycin plus ethambutol plus rifampin or rifabutin. Prophylaxis with azithromycin is recommended for patients with HIV and CD4 counts below 50.
5. *Treatment of Mycobacterium fortuitum*: *M. fortuitum*, which is commonly seen as an infection of central venous catheters or prosthetic devices is resistance to conventional antituberculous drugs. Surgical removal of the involved device is recommended and the agent is treated with regimens including amikacin and cefoxitin followed by long-term doxycycline or trimethoprim/sulfamethoxazole depending on in vitro sensitivities.
6. *Treatment of Mycobacterium chelonae*: *M. chelonae* has been treated with clarithromycin and azithromycin, but in serious infections amikacin and imipenem or cefoxitin may be necessary.
7. *Other mycobacterial species*: Specific regimens and controlled studies of treatments are not available for treatment of most of these diseases. Many of them are sensitive to a variety of antibiotics. This includes *M. marinum* which can be treated with a number of agents including rifampin, rifabutin, ethambutol, clarithromycin, and trimethoprim/sulfamethoxazole. Many strains are also sensitive to doxycycline and both clarithromycin and doxycycline have been used for treatment of superficial infections. Combinations of drugs are usually recommended for deep tissue infections.

Further Reading

Sia, I. G., & Wieland, M. L. (2011). Current concepts in the management of tuberculosis. Mayo Clin Proc, 86(4), 348–361.

Chapter 19
Clinical Approach to Treatment of Fungal Infections

Overview

The approach to treatment of fungal infections has focused on drugs that target the differences between the mammalian and the fungal cell membranes. The first antifungal drugs available commercially, the polyenes, bound to the ergosterol, the major fungal cell lipid (a lipid resembling mammalian cholesterol) achieving fungal killing by poking holes in the fungal cells. The azole drugs target ergosterol synthesis while the echinocandins take advantage of a unique fungal enzyme (1,3 D-glucan synthase) that is essential to cell wall integrity.

Drugs Used for Treatment of Fungal Infections

Polyene Antifungals

Nystatin, a polyene that contains multiple double bonds (Fig. 19.1) is an antifungal drug derived in 1950 from *Streptomyces noursei*. It has specificity for fungi based on its ability to bind specifically to ergosterol (the major fungal cell lipid, analogous to cholesterol in mammalian cells). It has a broad spectrum of activity against fungi but its toxicity when administered systemically has limited its use to topical treatments. It is commonly used to treat superficial oral or vaginal infections.

For many years, the only anti-infective agent licensed for use for systemic fungal infections was amphotericin B (Fig. 19.2). This agent, a polyene macrolide (characterized by many double bonds and a lactone ring) binds to the fungal lipid ergosterol in the fungal cell and creates pores that disturb the integrity of the membrane leading to cell death. As most fungal cells use ergosterol in the membranes, amphotericin B has a broad spectrum of activity. Certain exceptions to this rule include *Pseudallescheria* and *Candida lusitaniae*, both of which are intrinsically resistant to amphotericin B.

R.W. Finberg and R. Guharoy, *Clinical Use of Anti-infective Agents: A Guide on How to Prescribe Drugs Used to Treat Infections*,
DOI 10.1007/978-1-4614-1068-3_19,

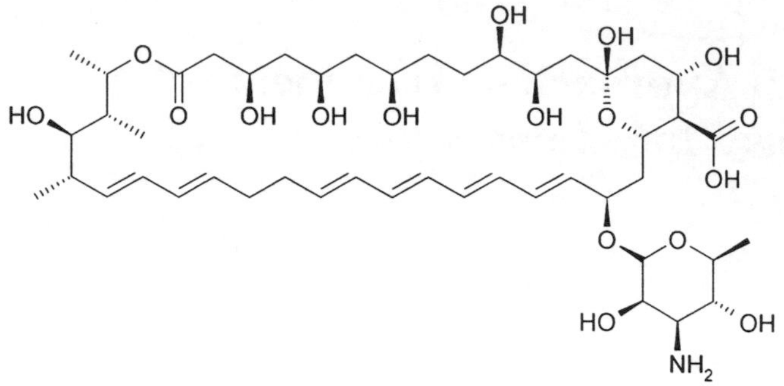

Fig. 19.1 Structure of nystatin

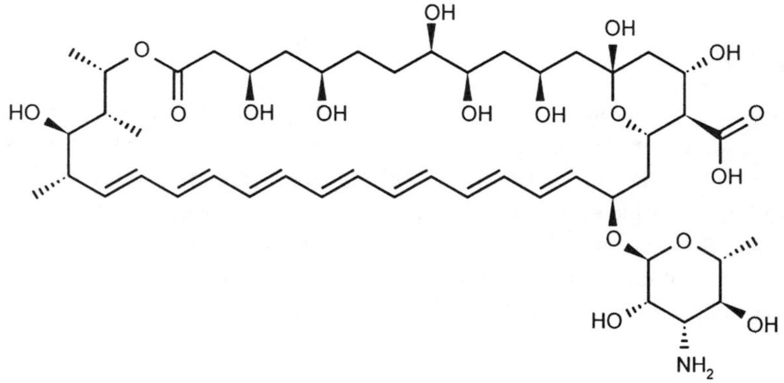

Fig. 19.2 Structure of amphotericin B

Amphotericin B is relatively insoluble and the administration of the drug in the usual phosphate buffer is associated with cytokine release leading to fever and rigors as well as renal toxicity. For this reason, the drug is routinely administered either as a liposomal formulation or in a lipid complex. These formulations have no effect on the drug's spectrum or antifungal activity but they do decrease toxicity. The drug is poorly absorbed and not available in any oral formulations. Unlike nystatin, it is not routinely used for treatment of local disease, although it has been used for bladder irrigation and gut "sterilization" protocols.

Azole Antifungals

The azole drugs revolutionized the treatment of fungi by making available oral drugs with few side effects that were still effective in the treatment of systemic infections. These drugs inhibit the synthesis of ergosterol, and therefore, in principle,

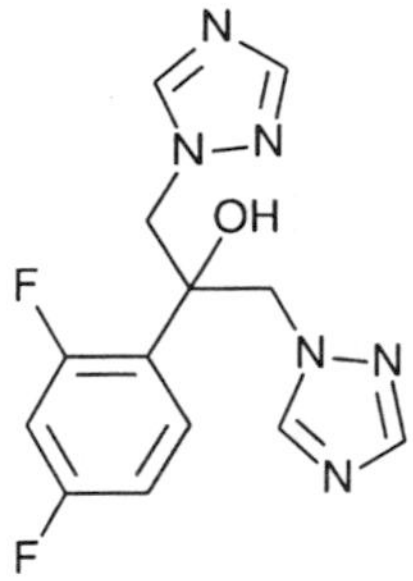

Fig. 19.3 Structure of fluconazole

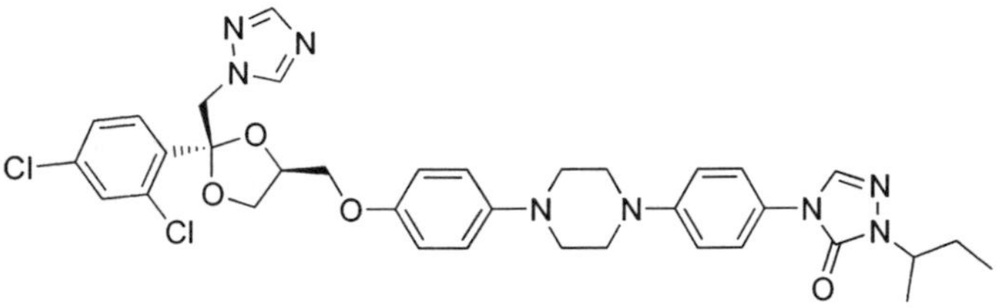

Fig. 19.4 Structure of itraconazole

could antagonize the action of amphotericin B, which functions by binding ergosterol. The different azoles vary in their spectrum with the original azole, ketoconazole, now having limited usefulness. Currently used azoles include fluconazole, itraconazole, voriconazole, and posiconazole. The major differences relate to spectrum and oral and parenteral administration (see Appendices).

Fluconazole is an exceptionally easy drug to use as it is available in both oral and IV formulations and is well absorbed and rapidly distributed throughout the body. Unlike amphotericin B, it is well concentrated in the urine and it crosses into the CSF (Fig. 19.3).

Itraconazole has a broader spectrum of activity than fluconazole, extending the spectrum to *Aspergillus* species in addition to having activity against non-*albicans Candida* spp. Absorption of this drug is less reliable than with fluconazole and it stimulates the hepatic p450 cytochrome system making it difficult to give to people taking calcineurin inhibitors (cyclosporine, tacrolimus, etc., are often used in the same patient population that is likely to be subject to fungal infections). Because of its broad spectrum of activity, this drug is often used to treat onychomycosis (fungal infections of the nails). Such treatment requires several weeks and is often complicated by the development of liver function abnormalities related to the drug (Fig. 19.4).

Voriconazole, a broad-spectrum agent with both PO and IV formulations, is the drug of choice for *Aspergillus* infections (Fig. 19.5). It is well absorbed and has sufficient penetration to the CSF (when given IV) that it can be used to treat CNS

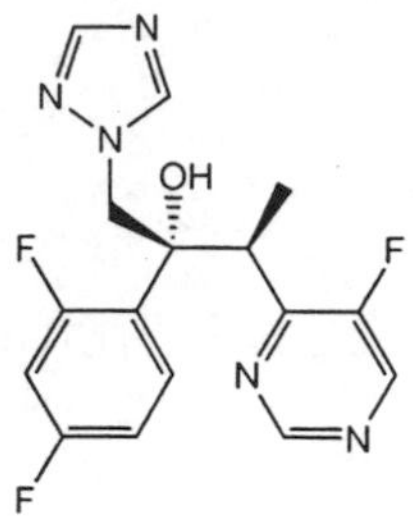

Fig. 19.5 Structure of voriconazole

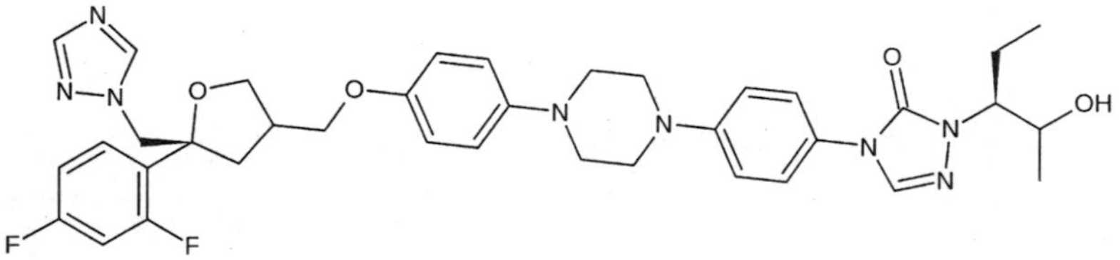

Fig. 19.6 Structure of posaconazole

Aspergillus infections. The major limitation of this drug is its lack of activity for *Mucor* species. Although posaconazole (Fig. 19.6) may have activity against some of these organisms, the treatment of choice for *Zygomycetes* remains amphotericin B.

Posaconazole, a PO azole with a very broad spectrum of activity has been successfully employed prophylactically in the prevention of fungal infection in stem cell transplant recipients. It can be used in the treatment of fungal infections that are resistant to fluconazole.

Echinocandins

The echinocandins kill fungi by inhibiting the enzyme 1,3-D-glucan synthase, an enzyme that is essential to the integrity of the fungal cell wall but is not required for mammalian cells. For this reason, the echinocandins (including caspofungin, micafungin, and anidulafungin) have little toxicity in humans. Their spectrum includes most *Candida* species including some *Candida* that are resistant to amphotericin B or azoles. They have less activity against *Candida parapsilosis* and *Candida guilliermondii*. These agents cannot be used to treat fungi that lack significant beta-glucan in their cell walls, such as *Cryptococcus neoformans, Trichosporon* species, and *Zygomycetes* organisms. On the other hand they have fungistatic activity against most *Aspergillus* species and can be used to treat *Aspergillus* (Fig. 19.7).

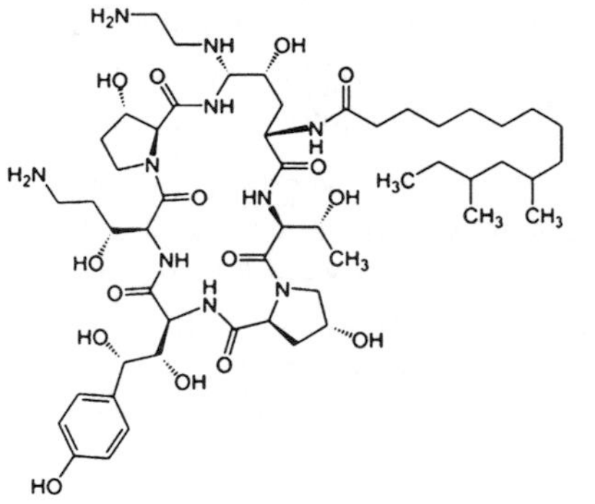

caspofungin

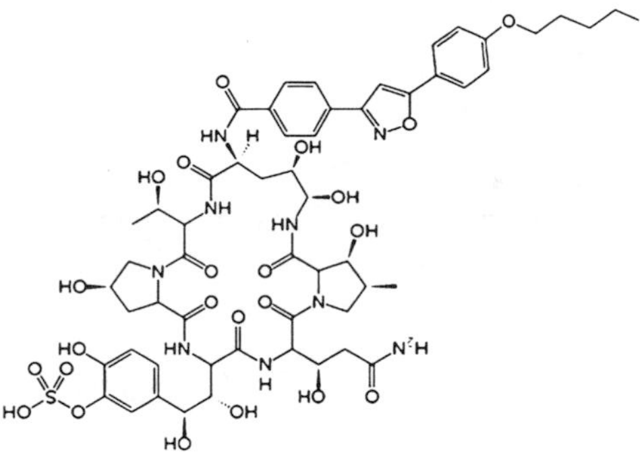

micafungin

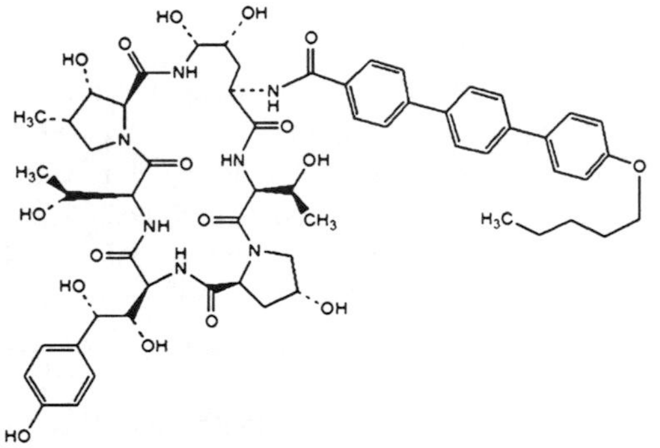

anidulafungin

Fig. 19.7 Structures of caspofungin, micafungin, and anidulafungin

Treatment of Specific Fungal Infections

1. **Candida species**:

   (a) *Candida albicans*: *Candida albicans*, the most common species of *Candida* seen clinically, is rapidly identified in the laboratory by the presence of a "germ tube." Most strains of *C. albicans* are sensitive to fluconazole and therefore this is the first drug of choice.

   (b) *Candida tropicalis*: While many strains of *C. tropicalis* are sensitive to fluconazole they are less sensitive to fluconazole than the *albicans* species and individual sensitivity testing may be necessary to establish the optimal agent for treatment. Alternative agents include voriconazole, posiconazole, itraconazole, an echinocandin (caspofungin, micafungin, or anidulafungin), or amphotericin B.

(c) *Candida krusei*: *C. krusei* species are often seen in units where fluconazole is used extensively. These strains are resistant to fluconazole but may be sensitive to other azoles or to echinocandins or amphotericin B.

2. **Aspergillus**: *Aspergillus fumigatus* is the most common pathogen amongst the *Aspergillus* genus. The drug of choice for most *Aspergillus* infections is voriconazole. Alternative agents include an echinocandin or an amphotericin preparation.
3. **Blastomycosis**: Amphotericin B (now usually given as a liposomal preparation) is still the drug of choice for serious systemic blastomycosis. Itraconazole has activity against blastomycosis but it may be necessary to measure serum levels to assure adequate dosing.
4. **Coccidioidomycosis**: *C. immitis* primary pulmonary disease ("valley fever") is usually a self-limited disease in a normal host and treatment is not required.
5. **Cryptococcus**: Cryptococcal meningitis is traditionally treated with amphotericin B. The addition of the pyrimidine analog 5-flucytosine (5-FC), both decreases the dose of amphotericin necessary and enhances the cure rate. One of the advantages of 5-FC is that, unlike amphotericin, it crosses well into the CSF. The 5-FC should not be used alone because of the rapid development of resistance to the single agent. This is one of the best-studied examples of antimicrobial synergy. Fluconazole also has activity against *Cryptococcus* and can be used as a single agent. In patients with HIV, treatment is often started with amphotericin plus 5-FC and once a dramatic response has been achieved, the patient is switched to fluconazole, which can be given orally.
6. **Histoplasmosis**: Uncomplicated, acute pulmonary histoplasmosis in healthy adults is usually self-limited and does not require treatment. Chronic disease and disease in immunocompromised patients can be treated with itraconazole. Severe or life-threatening illness should be treated with an amphotericin B lipid formulation.
7. **Paracoccidioidomycosis**: Paracoccidioidomycosis (South American blastomycosis) is a common cause of death in patients with HIV-1 infection and is commonly treated with trimethoprim/sulfamethoxazole or itraconazole.
8. **Sporotrichosis**: Cutaneous sporotrichosis can be treated with itraconazole while pulmonary, disseminated, or meningeal disease is treated initially with an amphotericin B formulation.

Further Reading

Smith, J. A., & Kauffman, C. A. (2010). Recognition and prevention of nosocomial invasive fungal infections in the intensive care unit. Crit Care Med, 38(8 Suppl), S380-387.

Terrell, C. L. (1999). Antifungal agents. Part II. The azoles. Mayo Clin Proc, 74(1), 78–100.

Tuon, F. F., Amato, V. S., & Penteado Filho, S. R. (2009). Bladder irrigation with amphotericin B and fungal urinary tract infection--systematic review with meta-analysis. Int J Infect Dis, 13(6), 701–706.

Chapter 20
Clinical Approach to Treatment of Viral Infections

In the last 20 years, we have begun to see some progress in the treatment of viruses. The treatment of viral infections other than HIV-1 has not reached the level of sophistication that treatment of bacterial infection has. Therefore, we have devoted less time to this subject than to the treatment of bacteria. The treatment of HIV-1 is a subject for a separate textbook, so in this book we are only presenting basic principles of treatment. The actual drug used changes dramatically from 1 year to the next based on availability of new agents as well as the development of resistance to some old drugs.

Most of the progress has been in the treatment of certain groups of viruses: herpes viruses, HIV, hepatitis B and C, and influenza.

1. *Treatment of herpes group viruses*: The herpes group of viruses includes herpes simplex types 1 and 2 (HSV-1 and HSV-2), varicella-zoster virus (VZV, the cause of chickenpox and shingles), cytomegalovirus (CMV), Epstein–Barr virus (EBV, the cause of infectious mononucleosis), Human Herpesvirus (HHV)-6 (the cause of roseola), HHV-7, and HHV-8 (Kaposi's sarcoma virus-KSHV).

   (a) *Acyclovir*: The discovery of acyclovir, or acycloguanosine, has had a profound effect on the treatment of herpes group viruses. This agent has been in use for over 20 years. Its primary use is in the treatment of herpes simplex encephalitis in normal hosts (a disease with a high morbidity and mortality if left untreated) or in prophylaxis and treatment of herpes simplex and VZV in immunocompromised patients (Fig. 20.1).

   Acyclovir, a purine analog, can be administered PO or IV. It has excellent activity against strains of HSV-1 and HSV-2, and the development of resistance is unusual, but has been reported in patients who have received multiple treatment courses. Acyclovir also has activity against VZV and is effective in treatment of disseminated disease in immunocompromised patients. Its short half-life of about 3 h means that it must be given several times a day when used orally. For this reason, valacyclovir (a prodrug of acyclovir which is better absorbed) is more commonly used when the

R.W. Finberg and R. Guharoy, *Clinical Use of Anti-infective Agents: A Guide on How to Prescribe Drugs Used to Treat Infections*,
DOI 10.1007/978-1-4614-1068-3_20,

Fig. 20.1 Structure of acyclovir

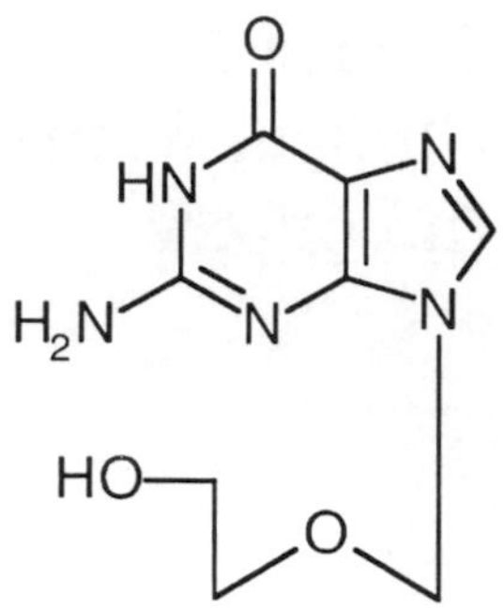

Fig. 20.2 Structure of ganciclovir

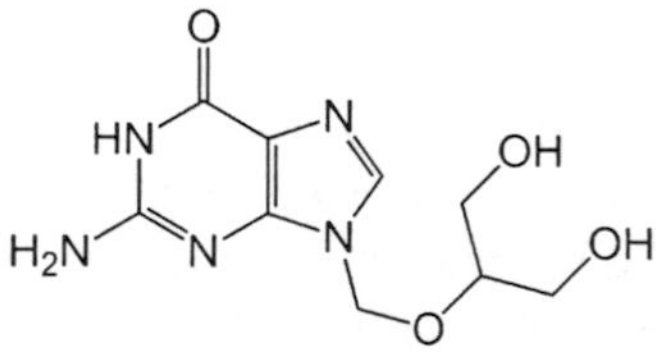

drug is given orally. These drugs can be given prophylactically to prevent reactivation of either HSV or VZV.

(b) *Ganciclovir*: Ganciclovir, in addition to having activity against HSV-1, HSV-2, and VZV, has activity against CMV (Fig. 20.2). This makes ganciclovir useful in the treatment of CMV infections in immunocompromised hosts including patients with HIV and bone marrow and solid organ transplant recipients who are sensitive to CMV reactivation and dissemination. Unlike acyclovir, ganciclovir has serious toxicity and a low toxicity/therapeutic ratio. Administration of ganciclovir is very commonly associated with neutropenia. The neutropenia associated with ganciclovir is usually reversible and may respond to cytokines such as G-CSF, but the increased incidence of bacterial infections in patients routinely treated with ganciclovir has led to its use only in the setting of documented CMV infection. Although the PO form of ganciclovir is poorly absorbed, valganciclovir can be given PO with more reliable absorption. CMV strains do become resistant to ganciclovir. Foscarnet (phosphonoacetic acid) and cidofovir can be used to treat ganciclovir-resistant CMV strains but foscarnet is difficult to administer, requiring IV infusions and constant monitoring of electrolytes.

(c) *Cidofovir*: Cidofovir, a polymerase inhibitor, has a broad spectrum of activity against DNA viruses as measured by in vitro sensitivity testing (Fig. 20.3). Its primary use, however, has been the treatment of ganciclovir-resistant CMV. Small studies suggest that, as seen in vitro, the drug may have activity against adenovirus and BK virus in immunocompromised patients; rigorous studies have yet to demonstrate clear efficacy. Cidofovir

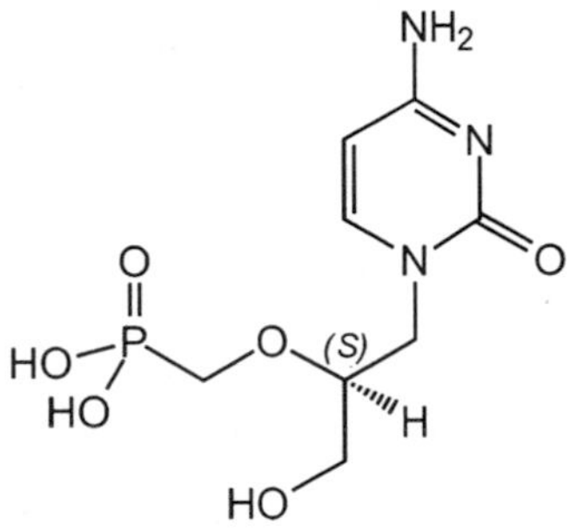

Fig. 20.3 Structure of cidofovir

administration is associated with nephrotoxicity and neutropenia. Because of its long half live, cidofovir can be administered weekly, making it an easy drug to give but this makes the side effects more long lasting.

2. *Treatment of influenza:*

   (a) *Amantadine and rimantadine*: The first discovered agents for treating influenza, amantadine and rimantadine, are active only against influenza A strains where they function by preventing uncoating of virus. In the past they were given orally and had efficacy in decreasing virus replication, and were effective when given prophylactically. However, their use in the elderly (natural targets for prophylactic administration) was somewhat limited by CNS side effects. They are no longer recommended for treatment of influenza as most strains are now resistant to these agents.

   (b) *Oseltamivir and zanamivir*: These agents, which are neuraminidase inhibitors, are active against both influenza A and B. Oseltamivir is given orally and has few side effects or toxicity making it an easy agent to administer. It can decrease virus shedding time when given early after acute infection in normal hosts, however resistance to the drug has been reported. Resistance to zanamivir may be less common, but use of this drug has been less widely accepted because of the need to administer it by inhaler.

3. *Treatment of hepatitis B and C*: Treatment of hepatitis B and C has evolved dramatically in the last few years and should continue to advance rapidly.

   (a) *Treatment of hepatitis B*: There is no currently accepted treatment for acute hepatitis B, which is most often a self-limited illness. Chronic hepatitis B, which can be associated with the development of cirrhosis and eventually fatal liver disease, has traditionally been treated with alpha-interferon (given subcutaneously). This human cytokine, produced as a recombinant protein in bacteria, is conjugated to polyethylene glycol (PEG) to assure higher serum levels and longer activity. The cytokine is biologically active, binding to specific receptors on many different cells. It is known to cause elevated temperatures and decreased neutrophil and platelet counts in normal individuals. Administration of this drug is often difficult because of its many toxic effects, which are by virtue of its mode of action (stimulation of

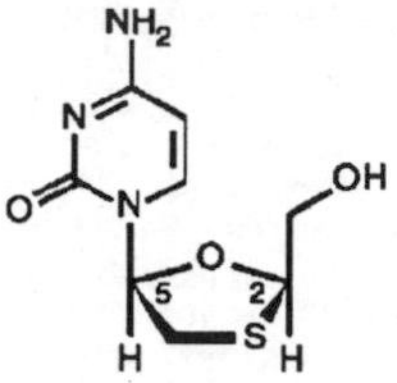

Lamivudine

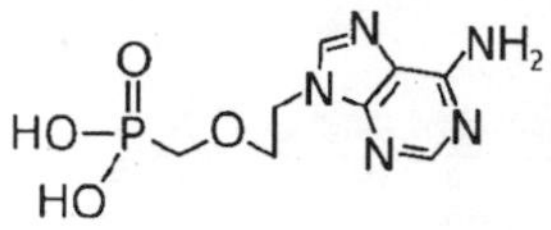

Adefovir

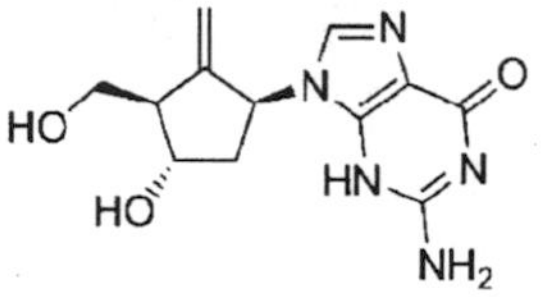

Entecavir

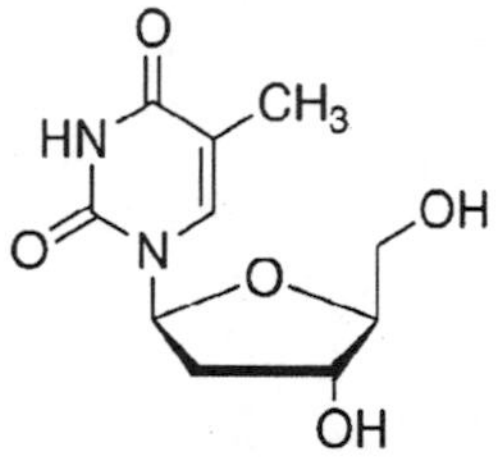

Telbivudine

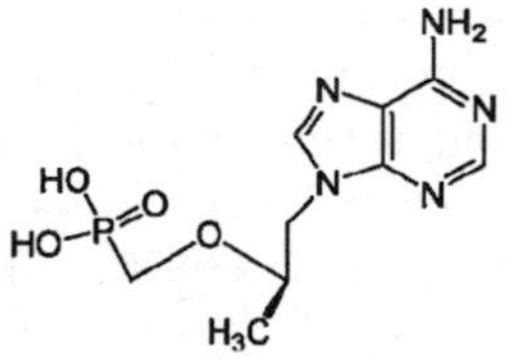

Tenofovir

Fig. 20.4 Structures of lamivudine, adefovir, entecavir, telbivudine, and tenofovir

host cells) very nonspecific. While the initial "flu-like symptoms" (fever, myalgias, and fatigue) that would be expected to occur after treatment with a cytokine usually disappear after the first days or weeks of therapy, the recommended 48-week course of therapy is often stopped prematurely because of intolerance from hematopoietic suppression, mood disorders, neuropathies, or profound fatigue. Recently several reverse transcriptase inhibitors have shown great efficacy with fewer side effects. Several nucleoside analogs, including lamivudine, adefovir, entecavir, telbivudine, and tenofovir, have all shown efficacy in the treatment of hepatitis B (Fig. 20.4). Side effects to these agents, all of which are administered orally, are much less common and they can usually be safely administered for a year or more. Although resistance has been documented to occur with lamivudine, adefovir, entecavir, and telbivudine, it has yet to be reported for the newest agent, tenofovir. However, optimistic projections need to be tempered by the short duration of clinical experience with the drug.

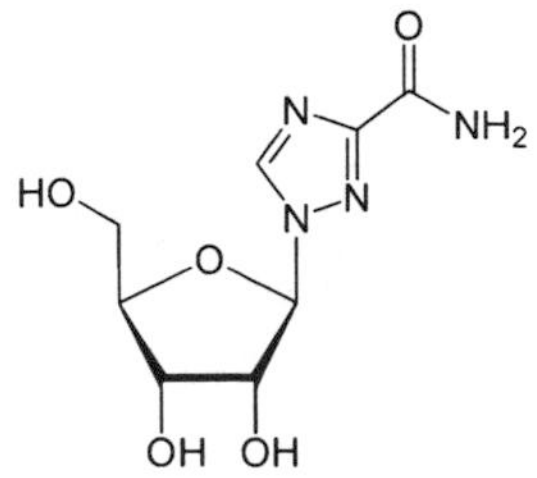

Fig. 20.5 Structure of ribavirin

(b) *Treatment of hepatitis C*: Treatment of chronic hepatitis C has traditionally focused on the use of interferon preparations in combination with the nucleoside analog ribavirin (Fig. 20.5).

Although the mechanism of action of ribavirin in the treatment of hepatitis C is not well defined (it has been attributed to ribavirin's effects on host cells as well as the ability of the drug to enhance the mutation rate in the virus), multiple studies reveal that the addition of ribavirin (given orally), to PEG-conjugated interferon (given subcutaneously), leads to better responses than treatment with interferon alone. The response of hepatitis C varies according to genotype, with 50% or less of patients with genotype 1 responding to treatment with 48 weeks of treatment with interferon plus ribavirin, while 75% of patients with genotype 2 may respond.

The licensing of two protease inhibitors, boceprevir and telaprevir, is anticipated to change the entire approach to treatment of hepatitis C and dramatically improve cure rates.

4. *Treatment of HIV*: In recent years there has been an explosion of options for treating HIV-1.
   There are five different classes of antiretrovirals:

1. Nucleoside reverse transcriptase inhibitors (NRTIs)
2. Nonnucleoside reverse transcriptase inhibitors (NNRTIs)
3. Protease inhibitors (PIs)
4. Integrase inhibitors (INSTI-integrase strand transfer inhibitor)
5. Cell entry inhibitors: fusion inhibitor, CRR5 antagonist.

As noted above, the treatment of HIV is a subject for a separate textbook. Since the drugs and resistance patterns change frequently, to find the latest recommendations, the reader is referred to the website maintained by the Department of Health and Human Services (DHHS). It can be reached at the following address: http://aidsinfo.nih.gov. Many of the drugs used for treatment of HIV-1 interact with other drugs (see Appendix G). Some general guidelines include the following:

1. Because of the ability of HIV-1 to rapidly develop resistance to any agent, multiple drugs should always be started at the same time. In general, this applies to stopping drugs as well. If one stops one drug without stopping the others, the development of resistance to the remaining agent(s) becomes extremely likely.

The half lives of certain drugs (like the NNRTIs) may be long and it is better to replace them with an alternative agent than to stop other drugs and leave the patient on NNRTI monotherapy as this will certainly lead to resistance.

2. The fewer pills the better. Adherence to therapy is important in terms of preventing the development of resistance. As single pill once a day is ideal. This has led to the development of drug combinations:

   (a) The most popular combinations at the time of publication of this book are the single pill Atripla, which consists of emtricitabine and tenofovir and efavirenz, or a combination of a protease inhibitor and two NRTIs.
   (b) Ritonavir is protease inhibitor (PI) that is also a P450 CYP3A4, 2D6 inhibitor that inhibits metabolism, and therefore boosts the plasma levels of other PIs. Therefore, this agent is commonly used in combination with other agents to maximize the efficacy of treatment and promote rapid decreases in viral loads.

      - Atazanavir can be used with ritonavir boosting (atazanavir/r) in combination with emtricitabine/tenofovir (Truvada). This results in a 3 pill regimen that can be taken once daily to enhance antiviral activity.
      - Darunavir with ritonavir boosting plus Truvada is a 4 pill daily regimen that produces an excellent viral response.

3. No documented "cures" of HIV-1 have been described. Therefore, once therapy is started it is anticipated to be lifelong. It is possible to bring viral loads down to undetectable levels but experience thus far has indicated that the virus will recur as soon as the drugs are stopped.
4. There is controversy about the best time to start antiviral therapy with some physicians starting therapy when CD4 counts are less than 500 while others wait until the CD4 count drops to less than 350. Responses in patients with CD4 counts less than 200 or with AIDS-defining illnesses are less likely and of shorter duration than for those with higher initial CD4 counts. Other reasons to start antiretroviral therapy (other than low CD4 count) include HIV-associated nephropathy, pregnancy, and the hepatitis B (treatment involves the use of overlapping drugs and therefore should be done at the same time).

Further Reading

"HIV/AIDS Basics | Questions and Answers | CDC HIV/AIDS." *Centers for Disease Control and Prevention*. Web. 09 Sept. 2011. <http://www.cdc.gov/hiv/resources/qa/definitions.htm>.

Ghany, M. G., Strader, D. B., Thomas, D. L., & Seeff, L. B. (2009). Diagnosis, management, and treatment of hepatitis C: an update. Hepatology, 49(4), 1335–1374.

Lok, A. S., & McMahon, B. J. (2009). Chronic hepatitis B: update 2009. Hepatology, 50(3), 661–662.

Part III
Sample Cases and Analyses of the Use of Anti-infective Agents

Chapter 21
Sample Cases and Analyses of the Use of Anti-infective Agents

Case 1: Streptococcal Sore Throat

An 8-year-old boy presents with a severe sore throat and fever. On physical examination the patient is noted to have an exudative pharyngitis and a fever of 40°C and some tender cervical adenopathy but is otherwise unremarkable. Lab results reveal a leukocytosis with a predominance of polymorphonuclear leukocytes and are otherwise within normal limits.

1. *What is the likely diagnosis?*

Answer: An exudative pharyngitis is likely to be caused by group A streptococci (*S. pyogenes*). This is an aerobic gram-positive coccus.

2. *Which of the following have activity against the causative organism?*
   (a) Penicillin
   (b) Cefazolin
   (c) Vancomycin
   (d) Azithromycin
   (e) Imipenem/cilistatin
   (f) Metronidizole
   (g) Levofloxacin
   (h) Amoxicillin
   (i) Piperacillin–tazobactam

Answer: The spectrum of metronidazole is limited to anaerobic organisms and some protozoans (including amoebae and *Giardia*). It does not have activity against group A streptococcus. All the other drugs listed have activity against aerobic gram-positive cocci and therefore could all be considered appropriate for treatment of this disease. Group A streptococcus has never developed resistance to penicillin and therefore penicillin is an acceptable choice for treatment. Penicillin can be given parenterally and a single dose of benzathine penicillin is all that is required.

R.W. Finberg and R. Guharoy, *Clinical Use of Anti-infective Agents: A Guide on How to Prescribe Drugs Used to Treat Infections*,
DOI 10.1007/978-1-4614-1068-3_21, © Springer Science+Business Media, LLC 2012

3. *What agent would you use? Why?*

Answer: Penicillin V is an acid-resistant formulation that is well absorbed and has efficacy in the prevention of rheumatic fever. On the basis of cost and efficacy, it is far and away the best choice of an oral antibiotic for treatment of streptococcal pharyngitis. The problem with the use of penicillin is that reliable bacterial eradication requires a 10-day course of a pill that must be taken 2–3 times a day. Because of its long tissue half-life, azithromycin can lead to much more rapid elimination of bacteria, but its ability to prevent rheumatic fever (largely a problem in children) is unproven. Since the major reason to treat streptococcal sore throats in children is the prevention of rheumatic fever, penicillin for a 10-day course is recommended. Treatment of streptococcal pharyngitis in adults is not usually required as the disease will resolve without any therapeutic intervention in most cases.

Case 2: Skin and Soft Tissue Infections

A 16-year-old boy presents with fever and redness around a skin break that occurred after a scrape in a football game.

1. *What are the likely organisms causing this disease?*

Answer: The likely organisms causing disease in this case are *Streptococcus* species (especially *S. pyogenes*) or *Staphylococcus* species.

2. *What anti-infective agent would you use to treat this disease?*

Answer: *S. pyogenes* produces toxins that cause rapid advancement into lymph and one can see red streaks that outline the lymph drainage of the extremity. This is known popularly in some cultures as "blood poisoning."

In this case, if we had a positive culture for *S. pyogenes* (definitive diagnosis), the patient could be treated with penicillin (IV or PO). Unfortunately, one often needs to start treatment without knowledge of the causative organism. In this case, the infection in the skin could also be caused by *Staphylococcus aureus* in addition to *S. pyogenes*. If this infection is caused by *S. aureus* it could be treated with a semi-synthetic penicillin (such as parenteral oxacillin or PO dicloxacillin) or cephalosporin (cephalexin is a commonly used oral agent). Unfortunately, with the rise of MRSA, which is commonly found in the community, appropriate treatment for all strains would involve vancomycin or an alternative agent with activity against gram-positive cocci (e.g., daptomycin or linezolid – see Chap. 8). Depending on the strain, some MRSA strains are sensitive to tetracyclines and TMP/SMX, and these drugs could be used; the choice should be guided on the resistance pattern of isolates found in the community in which the infection occurred.

Case 3: Uncomplicated Urinary Tract Infections

A healthy 25-year-old woman presents with a urinary tract infection. Such infections are most commonly caused by *E. coli*.

1. *Assuming this infection is caused by E. coli, which of the following antimicrobial agents would be appropriate treatment for this infection?*
   (a) Penicillin
   (b) Ampicillin
   (c) Cephalexin
   (d) Metronidazole
   (e) Azithromycin
   (f) Vancomycin
   (g) Trimethoprim–sulfamethoxazole
   (h) Ciprofloxacin
   (i) Piperacillin–tazobactam
   (j) Imipenem–cilistatin

Answer: Ampicillin, cephalexin, azithromycin, trimethoprim–sulfamethoxazole, ciprofloxacin, piperacillin–tazobactam, and imipenem–cilistatin, all have activity against *E. coli*.

2. *What agent would you chose? Why?*

Answer: Ampicillin, trimethoprim–sulfamethoxazole, cephalexin, and ciprofloxacin are all oral agents that are well concentrated in the urine and could be used to treat an infection in an outpatient. Although ampicillin was used for many years to treat *E. coli* infections, more recently because of its extensive use in most communities, many strains have become resistant. Trimethoprim–sulfamethoxazole or ciprofloxacin is now preferred even for treatment of patients not previously exposed to antibiotics. A variety of orally available cephalosporins can also be used. Many clinicians would begin with ciprofloxacin because of high rates of resistance to other agents in the community.

Case 4: A Complicated Urinary Tract Infection

A 68-year-old man with a history of repeated urinary tract infections associated with kidney stones and prostatic obstruction of the bladder presents to the hospital with fever and flank pain. His physical exam is remarkable only for fever and flank pain and his laboratory results include a leukocytosis (WBC of 18,000 mm^{-3}) and pyuria (200 cells in his urine). His previous history indicates that on his last hospitalization he was infected with an extended spectrum beta-lactamase (ESBL)-producing *Klebsiella*.

1. *How would you treat this patient?*

Answer: Organisms that produce ESBL are routinely resistant to both cephalosporins (including third-generation cephalosporins) and penicillins with conventional beta-lactamases (this includes piperacillin–tazobactam). While some ESBL-producing organisms may be sensitive to quinolones (such as ciprofloxacin), these gram negatives must commonly be treated with carbapenems (see Chap. 7). Therefore, he would be best treated with a carbapenem (e.g., ertapenem or meropenem).

Case 5: Community Acquired Pneumonia

A 45-year-old man presents with a high fever, chest pain, and a cough productive of greenish yellow sputum. Physical examination is remarkable for fever, tachypnea, and a chest exam that reveals signs of consolidation (egophony) in the left lower lobe. Laboratory examination is positive for an elevated WBC of 18,000 mm^{-3} with many immature forms seen on smear. Examination of the sputum reveals gram-positive cocci in pairs.

1. *What is the likely diagnosis?*

Answer: This is a classical presentation of pneumococcal pneumonia (pneumonia caused by *Streptococcus pneumoniae*) and such cases account for the majority of bacterial pneumonias. The key features that suggest bacterial pneumonia are the fever, leukocytosis, and productive cough. The presence of multiple gram-positive cocci in the sputum makes this the overwhelmingly likely diagnosis.

2. *Which of the following drugs would be considered effective treatment of this disease?*
   (a) Penicillin G
   (b) Ciprofloxacin
   (c) Oseltamivir
   (d) Vancomycin
   (e) Clindamycin
   (f) Aztreonam
   (g) Ceftriaxone

Answer: Penicillin G, clindamycin, ceftriaxone, and vancomycin all have activity against *S. pneumoniae*. Ciprofloxacin, oseltamivir, and aztreonam have little or no activity against *S. pneumoniae*. Although most strains of *S. pneumoniae* remain sensitive to penicillin, penicillin-resistant *S. pneumoniae* strains have been isolated in some communities. These exist in multiple forms with some strains being highly resistant. More commonly the strains are only moderately resistant and these strains can be successfully treated with ceftriaxone. Highly resistant strains are rare in most communities, but these would require vancomycin for treatment.

3. *How would you approach the patient who did not have "classical findings?"*

Answer: Although the patient described above is overwhelmingly likely to have *S. pneumoniae* disease, in many cases the etiology of the pneumonia is not clear and the differential diagnosis may include *Mycoplasma* and *Chlamydia*. In this case, an agent with activity against these intracellular organisms is indicated. Levofloxacin has activity against common respiratory pathogens including *S. pneumoniae*, as well as activity against intracellular organisms and can be used to treat community-acquired pneumonia. For seriously ill patients, requiring hospitalization, clinicians often prefer to use parenteral therapy with ceftriaxone. This agent has activity against moderately resistant *S. pneumoniae*, as well as a broad spectrum of gram-negative organisms. However, it does not have activity against *Mycoplasma* or *Chlamydia*, so it must be given in combination with another drug. It is usually combined with azithromycin.

4. *How would you treat a patient who develops pneumonia in the intensive care unit (ICU)?*

Answer: Because patients in hospitals, particularly ICUs, are likely to be colonized with gram-negative organisms such as *Pseudomonas aeruginosa*, as well as with MRSA, additional anti-infective agents are required. Since infection with *Mycoplasma* or *Chlamydia* is less likely to be of concern in this setting, a commonly used antimicrobial regimen would include piperacillin/tazobactam combined with vancomycin. Such a combination would have a broad spectrum of gram-negative activity and would include anaerobes that might have been acquired by aspiration as well as MRSA.

Case 6: A Patient with Meningitis

A 20-year-old college student presents with a 1-day history of high fever and severe headache. Physical examination is remarkable for a temp of 41°C and a rigid neck. The patient is responsive to questions but complains of a headache and stiff neck.

1. *What antimicrobial agent(s) should be given?*

Answer: Because of the possibility of resistant *S. pneumonia*e (the most common organism causing meningitis) vancomycin is routinely added to ceftriaxone as initial therapy. Ceftriaxone, which has excellent CSF penetration, has activity against *S. pneumonia*e, *N. meningitidis*, and *H. influenzae*.

2. *If this patient was older than 50, immunocompromised, or pregnant, how would this change your approach to therapy?*

Answer: In an immunocompromised patient *Listeria monocytogenes* is a common cause of bacterial meningitis. To treat *L. monocytogenes*, ampicillin should be added to ceftriaxone and vancomycin.

Case 7: A Patient with an Abdominal Perforation

An 86-year-old man presents with diverticulitis (a disease associated with colonic wall weakness and pouches that extend from the colon), which has led to perforation of the colon. Physical examination is remarkable for a temp of 38.6°C, tachycardia (P 128), tachypnea (RR 28), hypotension (BP 80/50), and a rigid abdomen.

1. *How would you treat this patient?*

Answer: The colon is filled with billions of bacteria most of which are anaerobic. If the integrity of the colon is compromised, these bacteria will spill into the peritoneal cavity, an ordinarily sterile area, and cause serious disease. In this case one needs to treat both for a broad spectrum of gram-negative aerobes (including *Pseudomonas*) as well as for *Bacteroides fragilis* (the most common organism in the colon). Logical choices include the use of a carbapenem (e.g., imipenem–cilistatin), a penicillin with anti-*Pseudomonas* activity combined with a beta-lactamase (to provide activity against *B. fragilis*), like piperacillin/tazobactam, or a third- or fourth-generation cephalosporin (to provide broad-spectrum gram-negative activity) combined with metronidazole (to kill *B. fragilis*).

Case 8: A Patient with Endocarditis

A 50-year-old man with a history of a bicuspid aortic valve presents with a 14-day history of fever and fatigue. Physical examination is significant for a fever of 38.5°C and a new diastolic heart murmur. Examination of the extremities was positive for Osler's nodes (painful nodules on his extremities) and splinter hemorrhages.

Laboratory data was significant for multiple sets of blood cultures that are positive for viridans streptococci.

1. *What antibiotics would be appropriate treatment for this patient?*

Answer: For sensitive strains of viridans streptococci, the first-line drug would be intravenous penicillin G (12–18 million units/day). Alternative therapies include ceftriaxone or vancomycin in case of penicillin allergies.

2. *What antibiotics should not be used in this case?*

Answer: Bacteriostatic antibiotics (like metronidazole or clindamycin) should not be used. While evidence for superiority of bactericidal versus bacteriostatic antibiotics is usually scant, in endocarditis it is clear that bacteriostatic antibiotics (such as the macrolides) have no place. The classic example of this is in the treatment of enterococcus endocarditis, an organism that is inhibited but not killed by penicillin. In this case, an aminoglycoside (such as gentamicin) needs to be added to the penicillin in order to achieve a cure.

Case 9: A Patient with Neutropenia

A 23-year-old man presents with acute myelogenous leukemia (AML). He is treated with high dose cytotoxic chemotherapy which results in a prolonged period of neutropenia. Immediately after receiving the chemotherapy he is afebrile and his physical examination is unremarkable. His laboratory examination is remarkable for having a white blood cell count (WBC) of 500 mm^{-3} with an absolute neutrophil count of less than 50 mm^{-3}.

1. *What anti-infective agents would you give to this patient to prevent the development of a bacterial infection?*

Answer: In this setting, infections with aerobic gram-positive organisms from the skin (staphylococci and streptococci) and aerobic gram-negative organisms from the GI tract (including lactose fermenters like *E. coli*, *Enterobacter*, *Klebsiella*, as well as nonlactose fermenters like *Pseudomonas*) are common. Use of quinolones, such as levofloxacin with a broad spectrum of activity against gram-positive and gram-negative aerobes, has been demonstrated to decrease the number of infections seen in these patients.

2. *Two weeks into his course of neutropenia, the patient spikes a high fever. Physical examination at this time is remarkable only for a temperature of 39°C. How would you treat the patient?*

Answer: Patients with persistent neutropenia are subject to transient bacteremias from the gut (this is often exacerbated by the chemotherapy they receive, which kills the cells of the gut as well as the blood), and in the absence of neutrophils they develop fatal bacterial infections. Antimicrobial therapy in this setting should be directed to aerobic gram-positive cocci and gram-negative rods (including *Pseudomonas aeruginosa*). The third-generation cephalosporin, ceftazidime, which has excellent activity against *Pseudomonas* is often used in this setting. If physical examination suggests the presence of skin erythema around a catheter site (a common clinical problem in this setting), an agent with activity against gram-positive cocci on the skin including streptococci, *S. aureus,* and *S. non-aureus* strains should be added. Vancomycin is commonly used in this setting.

Case 10: A Patient with HIV-1 Infection

A 38-year-old man with an HIV-1 infection has a CD4 count less than 50. Physical examination is not remarkable.

1. *What is this patient likely to be susceptible to, and how would one attempt to prevent infection?*

Answer: Patients with HIV-1, like most immunocompromised patients are likely to be infected by organisms in their environment that do not cause disease in normal hosts. Specific organisms that are of concern include *Pneumocystis jirovecii* and *Mycobacterium avium-intracellulare*.

Trimethoprim–sulfamethoxazole is commonly used when CD4 falls below 200 to prevent growth of *Pneumocystis* which causes fatal pneumonia, and azithromycin is used when CD4 counts drop below 100 to prevent growth of *Mycobacterium avium-intracellulare.*

Further Reading

Altamimi, S., Khalil, A., Khalaiwi, K. A., Milner, R., Pusic, M. V., & Al Othman, M. A. (2009). Short versus standard duration antibiotic therapy for acute streptococcal pharyngitis in children. Cochrane Database Syst Rev(1), CD004872.

Baddour, L. M., Wilson, W. R., Bayer, A. S., Fowler, V. G., Jr., Bolger, A. F., Levison, M. E., et al. (2005). Infective endocarditis: diagnosis, antimicrobial therapy, and management of complications: a statement for healthcare professionals from the Committee on Rheumatic Fever, Endocarditis, and Kawasaki Disease, Council on Cardiovascular Disease in the Young, and the Councils on Clinical Cardiology, Stroke, and Cardiovascular Surgery and Anesthesia, American Heart Association: endorsed by the Infectious Diseases Society of America. Circulation, 111(23), e394-434.

Gerber, M. A., Baltimore, R. S., Eaton, C. B., Gewitz, M., Rowley, A. H., Shulman, S. T., et al. (2009). Prevention of rheumatic fever and diagnosis and treatment of acute Streptococcal pharyngitis: a scientific statement from the American Heart Association Rheumatic Fever, Endocarditis, and Kawasaki Disease Committee of the Council on Cardiovascular Disease in the Young, the Interdisciplinary Council on Functional Genomics and Translational Biology, and the Interdisciplinary Council on Quality of Care and Outcomes Research: endorsed by the American Academy of Pediatrics. Circulation, 119(11), 1541–1551.

Guidelines for the management of adults with hospital-acquired, ventilator-associated, and healthcare-associated pneumonia. (2005). Am J Respir Crit Care Med, 171(4), 388–416.

Zalmanovici Trestioreanu, A., Green, H., Paul, M., Yaphe, J., & Leibovici, L. (2010). Antimicrobial agents for treating uncomplicated urinary tract infection in women. Cochrane Database Syst Rev(10), CD007182.

Appendix A
Common Infectious Diseases and Empiric Antibiotic Recommendations for Adult ED and Inpatients at UMass Memorial Medical Center*

Site/infection	Preferred[a]	Alternatives[a]
Urinary/kidney		
Asymptomatic bacteriuria	No treatment is indicated unless patient is pregnant or will undergo invasive urologic intervention(s) (if pregnant consider amoxicillin, nitrofurantoin, TMP–SMX OK IN SECOND TRIMESTER PREGNANCY)	
Symptomatic cystitis, uncomplicated (women only)	**Ciprofloxacin** 250 mg PO q12h **OR** **Nitrofurantoin** 100 mg PO q6h × 5d (NOT in CrCl < 60 ml/min; for uncomplicated, females only) – consider macrobid for outpatients	**TMP/SULFA** 1DS tablet PO q12h (caution in elderly, renal insufficiency) **OR** **Cefpodoxime** 200 mg PO BID
Symptomatic cystitis, complicated (with anatomic abnormality, indwelling foley cath., recent instrumentation, men, diabetes/other immunosuppression)	**Ciprofloxacin** 500 PO q12h vs. 400 mg IV q12h **OR** **Ceftriaxone** 1 g IV q24h	**Gentamicin:** maximum of 3–5 mg/kg/day per pharmacy protocol
Acute pyelonephritis	**Ceftriaxone** 1–2 g IV q24h (not for Enterococcus)	**Gentamicin:** Maximum of 3–5 mg/kg/day per pharmacy protocol **OR** ***Below agents, only if susceptibility known*** **TMP/SULFA** 1DS tablet PO q12h (caution in elderly, renal insufficiency) **OR** **Ciprofloxacin** 500 mg PO q12h/400 mg IVq12h
Nosocomial UTI	**Ceftazidime** 1 g IV q8h	**Gentamicin:** maximum of 3–5 mg/kg/day per pharmacy protocol

(continued)

*The authors acknowledge the contributions of Gail Scully, MD, Jennifer Daly, MD, and Elizabeth Radigan, PharmD in putting together these recommendations for our medical center.

R.W. Finberg and R. Guharoy, *Clinical Use of Anti-infective Agents: A Guide on How to Prescribe Drugs Used to Treat Infections*,
DOI 10.1007/978-1-4614-1068-3, © Springer Science+Business Media, LLC 2012

Site/infection	Preferred[a]	Alternatives[a]
Abdomen		
Extra-biliary: intra-abdominal abscess, diverticulitis, appendicitis		
• **Mild–moderate, community-acquired**	**Ceftriaxone** 1–2 g IV q24h ***PLUS*** **Metronidazole** 500 mg IV/PO q8-q12h	If severe PCN/CEPH allergy: **Aztreonam**[b] 1 g IV q8 ***PLUS*** **Metronidazole** 500 mg IV q8h ***PLUS*** **Vancomycin** to target troughs of 10–15 mcg/ml
• **Severe, community-acquired** Severe: debilitated or immunocompromised host, severe organ dysfunction	**Cefepime** 2 g IV q8-12h **OR** **Ceftazidime** 2 g IV q8-12h ***PLUS*** **Metronidazole** 500 mg IV/PO q8-q12h	**Ertapenem**[b] 1 g IV q24h **OR** **Piperacillin–tazobactam** 3.375 g IV q8h EI (extended infusion) over 4 h **OR** If severe PCN/CEPH allergy: see above
• **Hospital -acquired**	**Piperacillin–tazobactam** 3.375 g IV q8h EI over 4 h. ***If gram stain reveals GPC in clusters add*** **Vancomycin** to target troughs of 10–15 mcg/ml	**Imipenem**[b] 500 mg IV q6h. ***If gram stain reveals GPC in clusters add*** **Vancomycin** **OR** if severe PCN/CEPH allergy: **see above**
If Biliary (Cholangitis, Cholecystitis)	As above but consider adding vancomycin to cephalosporin regimens for enterococcal coverage	
Clostridium difficile		
• **Mild–moderate (creat <1.5 × baseline AND WBC < 15 k)**	**Metronidazole** 500 mg PO q8h	**Metronidazole IV** for patients who cannot take PO
• **Severe (creat >1.5 × baseline OR WBC >15 k)**	**Vancomycin** 125 mg PO q6h	NA
• **Severe, complicated (with hypotension, shock, ileus[X], or megacolon[X])**	**Vancomycin** 500 mg PO q6h ***PLUS*** **Metronidazole** 500 mg IV q8h	[X]Severe, complicated *C. difficile* with ileus/toxic megacolon, may add **Vancomycin** PR (500 mg in 100 ml NS, instill PR q6hr)
Bacterial meningitis (Note: Administer steroids prior to antibiotics: Dexamethasone 0.15 mg/kg IV q6h × 2–4 days)		
• **Age 18–50**	**Ceftriaxone** 2 g IV q12h ***PLUS*** **Vancomycin** 15 mg/kg IV q8-12h to target troughs of 15–20 mcg/ml	
• **Age >50, EtOH, immunocompromised host, pregnancy**	**Ampicillin** 2 g IV q4h ***PLUS*** **Ceftriaxone** 2 g IV q12h ***PLUS*** **Vancomycin** 15 mg/kg IV q8-12h to target troughs of 15–20 mcg/ml	

(continued)

Site/infection	Preferred[a]	Alternatives[a]
• **Post-neurosurgery, head trauma, VP shunt**	**Ceftazidime** 2 g IV q8h ***PLUS*** **Vancomycin** 15 mg/kg IV q8-12h to target troughs of 15–20 mcg/ml	**Meropenem**[b] 2 g IV q8h ***PLUS*** **Vancomycin**
Misc		
Fever in intravenous drug user (IDU)	**Vancomycin** 15 mg/kg IV q8-12h to target troughs of 10–20 mcg/ml ***PLUS*** **Ceftazidime** 2 g IV q8h	**Vancomycin *PLUS* Ciprofloxacin** 400 mg IV q12h
Septic shock in patient with IV catheter	**Vancomycin** 15 mg/kg IV q8-12h to target troughs of 10–20 mcg/ml ***PLUS*** **Ceftazidime** 2 g IV q8h	**Vancomycin *PLUS* Ciprofloxacin** 400 mg IV q12h
Skin/soft tissue/bone		
Bite wounds (human/animal)	**Ampicillin–sulbactam** 1.5 g IV q6h **OR** **Amoxicillin/clavulanate** 875 mg PO BID **OR** **Ceftriaxone** 1 g IV q24h ***PLUS*** **Metronidazole** 500 mg PO q8-12h	**Clindamycin** 300 mg PO QID ***PLUS*** **Ciprofloxacin** 500 mg PO BID
Cellulitis	Oral: **Cephalexin** 500 mg PO q6h (mild infections) **OR** **Dicloxacillin** 500 mg PO q6h (mild infections) IV: **Nafcillin** 2 g IV q4h **OR** **Cefazolin** 1–2 g IV q8h	**IV/PO: Clindamycin** (if PCN allergic, has good staph and strep coverage) 150-450 mg PO q6-8h or 600 mg IV q8h
	Suspicion MRSA (if open wound or pus, h/o MRSA): **Vancomycin** 15 mg/kg IV q12h to target troughs of 10–15 mcg/ml	Consider other antibiotics if MRSA is susceptible**: TMP/SULFA** 10 mg/kg/day of TMP IV or PO given in 2–4 divided doses **OR** **Doxycycline** 100 mg PO q12h
Diabetic foot infection (chronic ulcer with associated soft tissue infection)	**Vancomycin** to target troughs of 10–15 mcg/ml ***PLUS*** **Ceftazidime** 2 g IV q8h ***PLUS*** **Metronidazole** 500 mg IV/PO q8-12h	**Vancomycin *PLUS*** **Ciprofloxacin** 400 mg IV q12h ***PLUS*** **Metronidazole** 500 mg IV/PO q8-12h
Osteomyelitis (not diabetic foot, without associated soft tissue infection)	Empiric therapy NOT advised; HOLD antibiotics until cultures are obtained when possible	
Dry Gangrene only	Antimicrobials not indicated	

(continued)

Site/infection	Preferred[a]	Alternatives[a]
Lung		
Community-acquired pneumonia (CAP) includes patients coming from group homes and assisted living *FOR ICU admits, IV therapy for CAP*	**Azithromycin** 500 mg PO/IV q24 ***PLUS*** **Ceftriaxone** 1 g IV q24h (step down to PO cefpodoxime) For severe CAP or risk for MRSA: Consider adding **Vancomycin** 15 mg/kg IV q8-12h to target troughs of 15–20 mcg/ml **NOTE: If CAP patient is at risk for pseudomonas[c], consider: Piperacillin–tazobactam PLUS Ciprofloxacin**	Only if PCN/Ceph allergy: **Levofloxacin** 750 mg PO/IV q24h × 5 days OR 500 mg PO/IV × 7–10 days **NOTE: If CAP patient with PCN/CEPH allergy is at risk for pseudomonas[c], consider: Aztreonam[b]** 2 g IV q8h ***PLUS*** **Levofloxacin**
Early-onset HAP /VAP (within days 1–3 of admission)	As CAP above	
HCAP[d] and Late-onset HAP (day 4 or later): FLOOR admission	**Piperacillin/tazobactam** 4.5g IV q8h EI over 4 h ***OR*** **Ceftazidime** 2 g IV q8h (replaces pip/tazo) ***PLUS*** **Vancomycin** with target trough 15–20 mcg/ml **+/– Gentamicin** if risk for multi-drug resistant (MDR) pathogens (see risk below[e])	Obtain sputum cultures for gram stain, culture and sensitivity If severe PCN/CEPH allergy: **Aztreonam[b]** 2 g IV q8h (replaces pip/tazo) ***PLUS*** **Vancomycin** with target trough 15–20 mcg/ml
Late-onset VAP/HAP (day 4 or later): ICU admission	**Piperacillin/tazobactam** 4.5g IV q8h EI over 4 h ***OR*** **Ceftazidime** 2 g IV q8h (replaces pip/tazo) ***PLUS*** **Vancomycin** with target trough 15–20 mcg/ml **+/– Amikacin** if risk for MDR pathogens (see below[e])	Note: Amikacin does not require ID approval when used in ICUs

[a]**All doses are for patients with normal renal function**; for renal dosing consult UMass renal dosing handbook or call pharmacy

[b]Restricted antibiotic: requires ID approval

[c]Risk for Pseudomonas in CAP:) [COPD ***or*** Interstitial lung disease (e.g. pulmonary fibrosis) AND current *or* recent (within 3 months): corticosteroid (>10 mg prednisone daily) ***or*** antibiotic therapy ***or*** malnutrition] **OR** Structural lung disease (e.g. bronchiectasis)

[d]HCAP: Hospitalized for 2+ days in the past 90 days **OR** Resided in a nursing home or long term (acute) care facility (LTCF/LTAC) **OR** Attended hemodialysis/Received recurrent IV antibiotic therapy, chemotherapy or wound care in the past 30 days

[e]Risk factors for MDR organisms: Previous *Pseudomonas* pneumonia **OR** History of MDR organism only sensitive to aminoglycosides **OR** Received piperacillin–tazobactam/third or fourth generation cephalosporin/carbapenem in past 45 days **OR** Mechanical ventilation in past 30 days **OR** structural lung disease

[f]All recommendations are for empiric therapy pending microbiology results. They should not be used as a substitute for clinical judgment concerning individual patients. Note: For most indications it is appropriate to obtain two sets of blood cultures prior to beginning antibiotics; For

patients with severe sepsis of unknown origin (septic shock or multi-organ failure), an empiric antimicrobial regimen consisting of a dose of piperacillin/tazabactam (4.5 grams) and a dose of vancomycin (1gram) is reasonable, pending culture results and further work-up to establish a diagnosis. If there is a history of previous infection with MRSA, VRE, or ESBL organisms or any reason to suspect fungal disease (e.g. a history of prolonged neutropenia), alternative regimens should be considered, and a consultation with an expert in Infectious Diseases is recommended.

Appendix B
Bioavailability of Anti-infective Agents

Bioavailability of commonly used oral antimicrobials	
Acyclovir	15–30%
Amoxicillin	90%
Azithromycin	37%
Cefprozil	89–95%
Cefuroxime	37–52%
Cephalexin	90%
Ciprofloxacin	60–80%
Clindamycin	90%
Doxycycline	70–93%
Fluconazole	90%
Ganciclovir	6%
Levofloxacin	70–99%
Linezolid	100%
Metronidazole	100%
Minocycline	90%
Moxifloxacin	90–99%
Trimethoprim–Sulfamethoxazole	98–99%
Valacyclovir	55%
Valganciclovir	61% (Ganciclovir)
Voriconazole	65–96%

Bioavailability is a pharmacokinetic parameter that provides information on how much of an oral drug reaches the systemic circulation. This can vary due to differences in the rate and extent of dissolution and absorption of the oral form. Some commercially available parenteral agents are unsuitable for oral use due to poor absorption and bioavailability. Agents such as oral ciprofloxacin may not achieve serum or tissue concentrations equal to those achieved by the intravenous formulation, but they achieve adequate therapeutic concentrations to treat a specific infection. Agents such as oral acyclovir require higher dosages as compared to the intravenous formulation to reach adequate serum levels because of their low bioavailability. Azithromycin, on the other hand, has low bioavailability, but it is very well distributed to tissues making it suitable for treatment of many infections. Agents like oral nitrofurantoin concentrate in the urine and achieve therapeutic urine concentration making it a suitable agent for treatment of urinary tract infection. However, it is useless for treatment of systemic infections since it cannot achieve therapeutic serum levels. Some oral agents offer ≥90% bioavailability making them suitable for treatment of non-life-threatening and serious systemic infections

Example of antimicrobial agents that can be used to "step down" from intravenous to oral therapy

Target pathogen	Intravenous agent	Oral agent
Viridans streptococci, non-group D streptococci	Penicillin G	Amoxicillin, cephalexin
MSSA	Nafcillin, oxacillin	Cephalexin
Pseudomonas aeruginosa	Aminoglycosides, imipenem, meropenem, ceftazidime, cefepime	Ciprofloxacin
MRSA	Vancomycin, daptomycin, linezolid	Linezolid, doxycycline, trimethoprim/ sulfamethoxazole
VRE	Daptomycin, tigecycline, linezolid	Doxycycline, linezolid
Bacteroides fragilis	Clindamycin	Clindamycin, metronidazole
Candida albicans	Fluconazole, voriconazole, micafungin	Fluconazole, voriconazole, posaconazole
Aspergillus fumigatus	Voriconazole, micafungin, caspofungin	Voriconazole, posaconazole

Further Reading

Cunha, B. A. (2006). Oral antibiotic therapy of serious systemic infections. Med Clin North Am, 90(6), 1197–1222.

Lacy CF, Armstrong LL, Goldman MP, Lance LL. Drug Information Handbook. 18th ed. Hudson, OH: Lexi-Comp, Inc; 2009.

Micromedex Healthcare Series (intranet database). version 5.1. Greenwood village, Colo: Thompson Reuters (healthcare) Inc.

Appendix C
Dosage and Intervals for Administration of IV Antimicrobials

Class	Drug	Normal dose	Adjustment for impaired renal function (CrCl/min)		CRRT, hemodialysis, CAPD
			10–50	<10	
Adult patients					
Cephalosporins	Cefazolin	1–2 g q8h	q12h	q24h	CRRT: 1–2 g q12h HD: q24h (dose after HD on dialysis days) CAPD: 0.5 g q12h
	Cefepime	1–2 g q8h	q12h	1 g q24h	CRRT: 1–2 g q12h HD: 1 g q24h (after HD on dialysis days) CAPD: q48h
	Ceftaroline	600 mg q12h	30–50: 400 mg q12h 15–30: 300 mg q12h <15: 200 mg q12h		HD: 200 mg q12h (dose after dialysis on dialysis days) CRRT/PD: No data
	Ceftazidime	1–2 g q8h	q12h q24h		CRRT: q12-24h HD: after dialysis CAPD: 0.5 g q24h
	Ceftizoxime Cefotaxime	1–2 g q8h	q12h	q24h	CRRT: q 12h HD: q24h (after HD on dialysis days CAPD: 1 g q24h
	Cefoxitin	1–2 g q6-8h	q8-12h	q24-48h	CRRT: q8-12h HD/CAPD: q24h (after HD on dialysis days)
	Cefuroxime	0.75–1.5 g q8h	q12h	q24h	CRRT: q12h HD: after dialysis CAPD: q24h

(continued)

Class	Drug	Normal dose	Adjustment for impaired renal function (CrCl/min)		CRRT, hemodialysis, CAPD
			10–50	<10	
Fluoroquinolones	Ciprofloxacin	400 mg q12h	>30: q12h <30: q24h		CRRT: q12-24h HD: after dialysis CAPD: 200 mg q8h
	Gatifloxacin	400 mg q24h	LD 400 mg 200 mg qd	LD 400 200 mg qd	CRRT: 400 mg OD, HD/CAPD: 200 mg q24h (after HD on dialysis days)
	Levofloxacin	500–750 mg q24h	20–49: 750 q48h <20: 750 mg LD; 500 mg q48h		CRRT: 500 mg X1 then 250 mg q24h HD/CAPD: 750 mgX1; then 500 mg q48h
Carbapenems	Doripenem	500 mg q8h	30–50: 250 mg q8h 10–30: 250 mg q12h <10: no data		No data
	Ertapenem	1 g q24h	<30: 0.5 q24h		HD/CAPD: 0.5 g q24h CRRT: 1 g q24h
	Imipenem	0.5 g q6h	30–49: q8h 10–29: q12h <10: LD 500 mg, 250 mg q12h		CRRT: 500 mg q8h HD: 500 mg LD; 250 mg q12h CAPD: 250 mg q12h
	Meropenem	1 g q8h	q12h	q24h	CRRT: 1 g q12h HD: after dialysis CAPD: q24h
Macrolides	Azithromycin	240–500 mg q24h	No adjustment needed		
	Erythromycin	250–500 mg q6h	<10: 50–75% of dose		CRRT: Dose as for CrCl <10 HD/PD: None
Penicillins	Ampicillin	500 mg to 2 g q6h	q6-12h	q12-24h	CRRT: q6-12h HD/CAPD: q12h
	Ampicillin/ sulbactam	2 g/1 g q6h	q8-12h	q24h	CRRT: 1.5/0.75 q12h HD: after dialysis CAPD: 2g/1g q24h
	Penicillin G	1–4 million Units q6h	30-50: q6h 10–29: q8h <10: q12h		CRRT: q6-8h HD/CAPD: q12h (after HD on dialysis days)
	Piperacillin	3–4 g q6h	q6–8h	q8h	CRRT: q6-8h HD/CAPD: q8h (after HD on dialysis days)
	Piperacillin/ tazobactam	3.375–4.5 g q6h	2.25 q6	2.25 q8h	CRRT: 3.375 q8h HD: 2.25 q8h CAPD: 4.5 g q12h
	Ticarcillin/ clavulanate	3.1 g q4h	q8-12h	2 g q12h	CRRT: 3.1 g q8-12h HD: 2 g q12h CAPD: 3.1 g q12h

(continued)

Class	Drug	Normal dose	Adjustment for impaired renal function (CrCl/min)		CRRT, hemodialysis, CAPD
			10–50	<10	
Monobactams	Aztreonam	1–2 g q6-8h	10–30: 500 mg to 1 g q6-8h <10: 250–500 mg q6-8h		CRRT: q6-8h HD/CAPD: 500 mg q6-8h (after HD on dialysis days)
Miscellaneous	Colistin	1–1.5 mg/kg q8h	q24h	q36h	CRRT: 2.5 mg/kg q24h HD: 1.5 mg/kg after dialysis
	Daptomycin	4–6 mg/kg/day	<30: 4–6 mg q48h		HD/CAPD: 4–6 mg/kg q48h CRRT: 8 mg/kg q48h
	Linezolid	600 mg q12h	No adjustment needed		
	Tigecycline	50 mg q12h	No adjustment needed		
Aminoglycosides	Amikacin (traditional dosing)	5–7.5 mg/kg q8h	30–50: q12-18h; 10–29: q18-24h; <10: q48-72h		CRRT: 5–7.5 mg/kg q12-24h HD/CAPD: 5–7.5 mg/kg (after HD on dialysis days)
	Gentamicin, tobramycin (traditional dosing)	1–2 mg/kg q8h	30–50: q12-18h; 10–29: q18-24h; <10: q48-72h		CRRT: 1–2 mg/kg q12–24h HD/CAPD: 2 mg/kg (after HD on dialysis days)
Pediatric patients					
Cephalosporins	Cefazolin	25 mg/kg q8h	q12h	q24h	CRRT: 25 mg/kg q8h HD/CAPD: q24h (dose after HD on dialysis days)
	Cefepime	50 mg/kg q8-12h	q24h	q48h	CRRT: 50 mg/kg q12h HD/CAPD: 50 mg/kg q24h (after HD on dialysis days)
	Ceftazidime	25–50 mg/kg q8h	50 mg/kg q12-24h	50 mg/kg q48h	CRRT: 50 mg/kg q12h HD/CAPD: q48h (after HD on dialysis days)
	Ceftizoxime Cefotaxime	33–66 mg/kg q8h	q8-12h	q24h	CRRT: 50 mg/kg q8h HD/CAPD: q24h (after HD on dialysis days)
	Cefoxitin	20–40 mg/kg q6h	q8-12h	q24h	CRRT: 20–40 mg q8h HD/CAPD: 20–40 mg q24h (after HD on dialysis days)
	Cefuroxime	25–50 mg/kg q8h	30–50: normal dose 10–29: q12h <10: q24h		CRRT: 25–50 mg/kg q8h HD/CAPD: q24h (after dialysis on HD days)

(continued)

Class	Drug	Normal dose	Adjustment for impaired renal function (CrCl/min) 10–50	<10	CRRT, hemodialysis, CAPD
Fluoroquinolones	Ciprofloxacin	10–15 mg/kg q12h; CF 15 mg/kg q8-12h	30–50: normal dose 10–29: 10–15 mg/kg q18h <10: q24h		CRRT: 10–15 mg/kg q12h HD/CAPD: q24h (after dialysis on HD days)
Carbapenems	Doripenem	No data			
	Ertapenem	3 months to 12: 15 mg/kg q12h (max 1 g/day) ≥13: 1 g qd	No data		
	Imipenem	15–25 mg/kg q6h	7–13 mg/kg q8-12h	q24h	CRRT: 7–13 mg/kg q8h HD/CAPD: Q24h (after HD on dialysis days)
	Meropenem	20–40 mg/kg q8h	10–20 mg q12h	q24h	CRRT: 20–40 mg/kg q12h HD/CAPD: q24h (after HD on dialysis days)
Macrolides	Azithromycin	10 mg/kg q24h	No adjustments needed		
	Erythromycin	4–13 mg/kg q6h	Normal dose	q8h	CRRT: q6h HD/CAPD: q8h
Penicillins	Ampicillin	50 mg/kg q6h	q8-12h	q24h	CRRT: q6h HD/CAPD: q12h (after HD on dialysis days)
	Ampicillin/sulbactam	50 mg/kg (ampicillin) q6h	q8-12h	q24h	CRRT: q8h HD/CAPD: q24h (after HD on dialysis days)
	Penicillin G	50,000 units/kg q6h	30–50: q6h 10–29: q8h <10: q12h		CRRT: q8h HD/CAPD: q12h (after HD on dialysis days)
	Piperacillin	50–75 mg/kg q6h	q8-12h	q12h	CRRT: q8h HD/CAPD: q8h (after HD on dialysis days)
	Piperacillin/tazobactam	50–75 mg/kg q6h	35–50 mg q6-8h	q8h	CRRT: 35–50 mg/kg q8h HD/CAPD: 50–75 mg/kg q12h (after HD on dialysis days)
	Ticarcillin/clavulanate	50–75 mg/kg q6h	30–50: normal dose 10–29: q8h <10: q12h		CRRT: 50–75 mg/kg q8h HD/CAPD: q12h (after HD on dialysis days)
	Nafcillin Oxacillin	1–2 g q6h	No adjustment needed		

(continued)

Class	Drug	Normal dose	Adjustment for impaired renal function (CrCl/min) 10–50	<10	CRRT, hemodialysis, CAPD
Monobactams	Aztreonam	30–40 mg/kg q8h	30–50: normal dose 10–29: 15–20 mg/kg q8h <10: 7.5–10 mg/kg q12h		CRRT: 30–40 mg/kg8h HD/CAPD: 7.5–10 mg/kg (after HD on dialysis days)
Miscellaneous	Daptomycin	6 mg/kg q24h	10–29: 4 mg/kg q24h <10: q48h		CRRT: 8 mg/kg q48h HD/CAPD: 4 mg q48h (after HD on dialysis days)
	Linezolid	600 mg q12h	No adjustment needed		
	Tigecycline	50 mg q12h	No adjustment needed		
Amino-glycosides	Amikacin	5–7.5 mg/kg q8h	30–50: q12-18h; 10–29: q18-24h; <10: q48-72h		CRRT: 7.5 mg/kg q12-24h HD/CAPD: 5 mg/kg (after HD on dialysis days)
	Gentamicin, tobramycin (traditional dosing)	2.5 mg/kg q8h	30–50: q12-18h; 10–29: q18-24h; <10: q48-72h		CRRT: 2–2.5 mg/kg q 12-24h HD/CAPD: 2 mg/kg (after HD on dialysis days)

Further Reading

Bennett, W. M., Aronoff, G. R., Morrison, G., Golper, T. A., Pulliam, J., Wolfson, M., et al. (1983). Drug prescribing in renal failure: dosing guidelines for adults. Am J Kidney Dis, 3(3), 155–193.

Eyler, R. F., & Mueller, B. A. (2010). Antibiotic pharmacokinetic and pharmacodynamic considerations in patients with kidney disease. Adv Chronic Kidney Dis, 17(5), 392–403.

Patel, N., Scheetz, M. H., Drusano, G. L., & Lodise, T. P. (2010). Determination of antibiotic dosage adjustments in patients with renal impairment: elements for success. J Antimicrob Chemother, 65(11), 2285–2290.

Siberry, G. K., Iannone, R., & Harriet Lane Home (Baltimore Md.). (2000). *The Harriet Lane handbook : a manual for pediatric house officers* (15th ed.). St. Louis ; London: Mosby.

St Peter, W. L., Redic-Kill, K. A., & Halstenson, C. E. (1992). Clinical pharmacokinetics of antibiotics in patients with impaired renal function. Clin Pharmacokinet, 22(3), 169–210.

Appendix D
Dosage and Intervals for Administration of Oral Antimicrobials

Class	Drug	Strength	Usual adult dose	Usual pediatric dose
Cephalo-sporins	Cefaclor	250, 500 mg	500 mg q8h	>1 month: 20–40 mg/kg/day divided q8-12 h (max. 2 g/days)
	Cefadroxil	500 mg, 1 g	1–2 g in 1–2 divided doses	30 mg/kg/day divided twice daily (max. 2 g/day) Adolescents: 1–2 g/day in divided doses
	Cefpodoxime	100, 200 mg	200 mg q12h	>6 months–12 years: 10 mg/kg/day divided every 12 h (max. 800 mg/day) Adolescents: 100–400 mg dose every 12 h
	Ceftibuten	400 mg	400 mg daily	To <12 years: 9 mg/kg/day (max. 400 mg/day) Adolescents: 400 mg/day
	Cefprozil	250, 500 mg	500 mg q12h	>6 months to 12 years: 15–30 mg/kg/day divided every 12 h (max. 1 g/day) Adolescents: 250–500 mg every 12 h
	Cefuroxime	125, 250, 500 mg	500 mg bid	>3 months to 12 years: 20–30 mg/kg/day divided every 12 h (max. 1 g/day) Adolescents: 250–500 mg twice daily
	Cephalexin	250, 500 mg	500 mg q6h	Children: 25–100 mg/kg/day divided every 6–8 h (max. 4 g/day)

(continued)

Class	Drug	Strength	Usual adult dose	Usual pediatric dose
Fluoroquinolones	Ciprofloxaxin	250, 500, 750 mg	500 mg q12h	Children: 20–40 mg/kg/day divided every 12 h (max. 1.5 g/day)
		500, 1,000 mg ER	1,000 mg q24h	
	Gemifloxacin	320 mg	320 mg daily	
	Levofloxacin	250, 500 and 750 mg	500 mg daily	6 months to 5 years: 10 mg/kg every 12 h
				>5 years: 10 mg/kg/day (max. 500 mg/day)
	Moxifloxacin	400 mg	400 mg daily	
	Norfloxacin	400 mg	400 mg bid	
	Ofloxacin	200, 300, 400 mg	400 mg q12h	15 mg/kg/day every 12 h
Glycopeptide	Vancomycin	125, 250 mg	125 mg q6h	40 mg/kg/day in divided doses every 6 h (max. 2 g/day)
Macrolides	Azithromycin	250, 500 mg	500 mg day 1, then 250 mg days 2–5	<6 months: 10 mg/kg/day once daily
			2 g single dose	>6 months: 10–20 mg/kg/day (max. 2 g/day)
		2 g susp.		
	Clarithromycin	250, 500 mg	500 mg q 12h	15 mg/kg/day divided every 12 h
	Erythromycin	250, 333, 500 mg	500 mg q6h	Neonates: 20–40 mg/kg/day in divided doses every 6–12 h
				Infants and children: 30–50 mg/kg/day divided every 6–8 h (max. 2 g/day base)
Penicillins	Penicillin V	250, 500 mg	500 mg q6h	<12 years: 25–50 mg/kg/day divided every 6–8 h (max. 3 g/day)
				>12 years: 125–500 mg every 6–8 h
	Amoxicillin	250, 500 mg	500 mg q8h	Neonates: <3 months: 20–30 mg/kg/day in divided doses every 12 h
				>3 months and children: 25–50 mg/kg/day in divided doses every 8–12 h
	Amoxicillin/clavulanate	250/125; 500/125; 875/125	875 mg q12h	Neonates: <3 months: 30 mg/kg/day divided every 12 h (use 125 mg/5 ml suspension)

(continued)

Class	Drug	Strength	Usual adult dose	Usual pediatric dose
		1,000/62.5 ER	2,000 mg q12h	Children <40 kg: Amoxicillin 20–40 mg/kg/day in divided doses every 8 h
	Ampicillin	250, 500 mg	500 mg q6h	50–100 mg/kg/day divided every 6 h (max. 2–3 g/day)
	Dicloxacillin	125, 250, 500 mg	500 mg q6h	<40 kg: 25–50 mg/kg/day divided every 6 h (max. 2 g/day) >40 kg: 125–500 mg every 6 h (max. 2 g/day)
Oxazolidinone	Linezolid	600 mg	600 mg bid	Neonates: <12 years: 10 mg/kg/dose every 8–12 h >12 years and adolescents: 600 mg every 12 h
Tetracycline	Doxycycline	50, 100 mg	100 mg bid	>8 years: 2–4 mg/kg/day divided every 12 h (max. 200 mg/day) Adolescents: 100–200 mg/day in 1–2 divided doses
	Minocycline	50, 75, 100 mg	200 mg once and 100 mg bid	>8 years: 4 mg/kg loading dose followed by 2 mg/kg every 12 h >12 years: 50–100 mg once or twice daily
	Tetracycline	250, 500 mg	500 q6h	>8 years: 25–50 mg/kg/day in divided doses every 6 h (max. 3 g/day) Adolescents: 250–500 mg every 6–12 h
Miscellaneous	Fidaxomicin	200 mg	200 mg twice daily	
	Metronidazole	250, 500 mg	500 mg three times a day	Neonates: 7.5–30 mg/kg/day in divided doses every 12–48 h
		750 mg ER	750 mg daily	Children: 15–50 mg/kg/day in divided doses every 6–8 h
	Nitrofurantoin	25, 50, 100 mg 100 mg (monohydrate, macrocrystals)	100 mg q6h 100 mg q12h	1–7 mg/kg/day divided every 6 h (max. 400 mg daily)
	Trimethoprim–SMX	400/80 mg, 800/160 mg	1 (800/160 mg) qd-12h	>2 months: 6–20 trimethoprim mg/kg/day in divided doses every 6–12 h

Appendix E
Important Interactions with Commonly Used Anti-infectives

Antifungals – azole

Interacting drug (A)	Fluconazole (B)	Itraconazole (B)	Ketoconazole (B)	Posaconazole (B)	Voriconazole (B)	Effect
Calcium channel blockers	X	X	X		X	↑ Levels of A
Carbamazepine		X			X	↓ Levels of B
Hydantoins	X	X	X	X	X	↑ Levels of A, ↓ levels of B
Lovastatin/ simvastatin		X			X	↑ Levels of A: rhabdomyolysis reported
Midazolam/ triazolam	X	X	X	X	X	↑ Levels of A
Oral anticoagulants	X	X	X		X	↑ Levels of A
Oral hypoglycemics	X	X			X	↑ Levels of A
Pimozide				X	X	↑ Levels of A
Proton pump inhibitors		X	X	X	X	↑ Levels of A
Rifampin/ rifabutin	X	X	X	X	X	↑ Levels of A, ↓ levels of B
Sirolimus				X	X	↑ Levels of A
Tacrolimus	X		X	X	X	↑ Levels of A
Trazodone			X			↑ Levels of A

Antifungals – echinocandin: caspofungin (A)

Interacting drug (B)	Effect
Cyclosporine	↑ Levels of A
Tacrolimus	↓ Levels of B
Carbamazepine, dexamethasone, phenytoin, rifamycin	↓ Levels of A: ↑ caspofungin dose to 70 mg/day

Antibacterials

Agent (A)	Interacting drug (B)	Effect
Aminoglycosides	Amphotericin B	↑ Nephrotoxicity
	Loop diuretics	↑ Ototoxicity
Amphotericin products	Aminoglycosides, foscarnet, cidofovir, pentamidine, cyclosporine	↑ Nephrotoxicity
Ampicillin, Amoxicillin	Allopurinol	Increased frequency of rash
Atovaquone	Tetracycline	↓ Levels of A
Daptomycin	HMG-CoA reductase inhibitors	↑ Toxicity
Doripenem	Probenecid	↑ Levels of A
	Valproic acid	↓ Levels of B
Doxycycline	Warfarin	↑ Effects of B
Linezolid	Adrenergic agents	Risk of hypertension
	Rifampin	↓ Levels of A
	SSRIs	Serotonergic syndrome
Macrolides		
Erythromycin, azithromycin, clarithromycin	Pimozide	↑ Q-T interval
Erythromycin, clarithromycin	Carbamazepine	↑ Levels of B
	Colchicine	↑ Levels of B
	Ergot alkaloids	↑ Levels of B
	Lovastatin, simvastatin	↑ Levels of B
	Tacrolimus	↑ Levels of B
Metronidazole	Cyclosporine	↑ Levels of B
	Lithium	↑ Levels of B
	Oral anticoagulants	↑ Levels of B
	Phenobarbital, hydantoin	↑ Levels of B
Piperacillin/tazobactam	Methotrexate	↑ Levels of B
Quinupristin–dalfopristin	Paclitaxel, vincristine	↑ Levels of B
	Calcium channel blockers	↑ Levels of B
	Cyclosporine, tacrolimus	↑ Levels of B
	Midazolam, triazolam	↑ Levels of B
	Statins	↑ Levels of B

(continued)

Antibacterials (continued)

Agent (A)	Interacting drug (B)	Effect
Rifamycin (rifampin, rifabutin)	Caspofungin	↓ Levels of B
	Clarithromycin	↑ Levels of A, ↓ levels of B
	Cyclosporine	↓ Levels of B
	INH	Conversion to toxic hydrazine
	Itraconazole, ketoconazole	↓ Levels of B, ↑ levels of A
	Linezolid	↓ Levels of B
	Tacrolimus	↓ Levels of B
Tetracycline	Digoxin	↑ Toxicity of B
Trimethoprim–Sulfamethoxazole	ACE inhibitors	↑ Serum K^+ level
	Amantadine	↑ Levels of B
	Methotrexate	↑ Bone marrow depression

Quinolones

Interacting drug	Ciprofloxacin	Gemifloxacin	Levofloxacin	Moxifloxacin	Effects
Procainamide, amiodarone			X	X	↑ QT interval
Insulin, oral hypoglycemic	X	X	X	X	↑ and ↓ Blood sugar
Cations (Al+++, Ca++, Fe++, Mg++, Zn++)	X	X	X	X	↓ Absorption of quinolones
Methadone	X				↑ Levels of methadone
NSAIDs	X		X		↑ Risk seizures/ CNS stimulation

Further Reading

Hansten PD, Horn JR. (2010) *The top 100 drug interactions*. Washington, D.C.: H and H publications.

Micromedex ® Healthcare Series. Copyright 1974–2011. Thompson Healthcare

Appendix F
Oral Agents that May have Interactions with Food

Agent	Interaction	Management
Ampicillin	↓ Ampicillin concentration	Take 1 h before or 2 h after meals
Atovaquone	High fat ↑ atovaquone level	Take with meals
Cefaclor	May ↓ cefaclor level	Take on empty stomach
Ciprofloxacin, levofloxacin, moxifloxacin	Cations (iron, calcium, aluminum, magnesium) antacid, dairy products	Take 2 h before or after administration of cations or antacids or dairy products
Didanosine	↓ Didanosine concentration	Take on empty stomach
Efavirenz	↑ Absorption	Take on empty stomach
Entecavir	May ↓ entecavir level	Take on empty stomach
Erythromycin	Grapefruit juice: may ↑ bioavailability	Avoid grapefruit juice
Indinavir	↓ Bioavailability of indinavir	Take on empty stomach
Isoniazid	Tyramine containing food: may ↑ blood pressure	Avoid food with tyramine (sauerkraut, soy sauce, tap beers, dried meats or red wine)
	Food: may ↓isoniazid level	Take on empty stomach
	Histamine-containing food: ↑ histamine effect	Fish (tuna, skipjack or other tropical fish) may cause histamine related effects
Itraconazole	Food: may ↑ or ↓ bioavailability dependent on dosage form	Take capsule after eating a meal Take oral solution on empty stomach
	Grapefruit juice: ↓ bioavailability	Avoid grapefruit juice
Linezolid	Tyramine containing food: ↑ pressor response	Avoid tyramine containing foods
Minocycline	Cations, antacids, dairy products: ↓ minocycline level	Take 2 h before or after administration of cations or antacids or dairy products
Nelfinavir	↑ Nelfinavir level	Take with food or milk
Posaconazole	May ↑ posaconazole level	Take with food
Primaquine	Grapefruit juice: may ↑ level	Avoid grapefruit juice

(continued)

Agent	Interaction	Management
Rifampin	May ↓ rifampin level	Take on empty stomach, 1 hour before or 2 h after a meal
Saquinavir	Grapefruit juice: may ↑ concentration	Avoid grapefruit juice
Tetracycline	Cations, antacids, dairy products: ↓ tetracycline level	Take on empty stomach
Valganciclovir	High fat food may ↑ ganciclovir level	Take with food or milk
Voriconazole	Possible ↓ voriconazole level	Take on empty stomach

Further Reading

Koda-Kimble MA, Young LY, Kradjan WA, Guglielmo BJ, Alldredge BK, Corelli RL. (2007). *Handbook of Applied Therapeutics* (8th ed.). Lippincott, Williams and Wilkins.
Micromedex ® Healthcare Series. Copyright 1974–2011. Thompson Healthcare

Appendix G
Drug Interactions with Anti-retroviral Agents

Table G.1 Drug interactions between antiretrovirals and other drugs: protease inhibitors (PI)

PI	Other drug	Effect	Comments
Acid reducers			
ATV ± RTV	Antacids	↓ ATV expected when given simultaneously	Give ATV at least 2 h before or 1 h after antacids or buffered medications
FPV		APV AUC ↓ 18%, no significant change in APV C_{min}	Give FPV simultaneously with or at least 2 h before or 1 h after antacids
TPV/r		TPV AUC ↓ 27%	Give TPV at least 2 h before or 1 h after antacids
RTV-boosted PIs			
ATV/r	H2 receptor antagonists	↓ ATV	H_2 receptor antagonist dose should not exceed a dose equivalent to famotidine 40 mg BID in ART-naïve patients or 20 mg BID in ART-experienced patients Give ATV 300 mg + RTV 100 mg simultaneously with and/or 10 h after the H_2 receptor antagonist If using TDF and H_2 receptor antagonist in ART-experience patients, use ATV 400 mg + RTV 100 mg
DRV/r, LPV/r		No significant effect	
PIs without RTV			
ATV	H2 receptor antagonists	↓ ATV	H_2 receptor antagonist single dose should not exceed a dose equivalent of famotidine 20 mg or total daily dose equivalent of famotidine 20 mg BID in ART-naïve patients Give ATV at least 2 h before and at least 10 h after the H_2 receptor antagonist

(continued)

Table G.1 (continued)

PI	Other drug	Effect	Comments
FPV		APV AUC ↓ 30%, no significant change if APV C_{min}	Give FPV at least 2 h before H2 receptor antagonist if concomitant use is necessary. Consider boosting with RTV
ATV	Proton pump inhibitors (PPIs)	↓ ATV	**PPIs are not recommended in patients receiving unboosted ATV.** In these patients, consider alternative acid – reducing agents, RTV boosting, or alternative PIs
ATV/r		↓ ATV	PPIs should not exceed a dose equivalent to 0meprazole 20 mg daily in PI-naive patients. PPIs should be administered at least 12 h prior to ATV/r **PPIs are not recommended in PI-experienced patients**
DRV/r, TPV/r		↓ Omeprazole PI: no significant effect	May need to increase omeprazole dose with TPV/r
FPV ± RTV, LPV/r		No significant effect	
SQV/r		SQV AUC ↑ 82%	Monitor for SQV toxicities
Anticoagulants			
ATV ± RTV, DRV/r, FPV ± RTV, LPV/r, SQV/r, TPV/r	Warfarin	↑ or ↓ Warfarin possible DRV/r ↓ S-warfarin AUC 21%	Monitor INR closely when stopping or starting PI and adjust warfarin dose accordingly
Anticonvulsants			
RTV-boosted PIs			
ATV/r, FPV/r, LPV/r, SQV/r, TPV/r	Carbamazepine	↑ Carbamazepine possible TPV/r ↑ carbamazepine AUC 26% may ↓ PI levels substantially	Consider alternative anticonvulsant or monitor levels of both drugs and assess virologic response. **Do not coadminister with LPV/r once daily**
DRV/r		Carbamazepine AUC ↑ 45% DRV: no significant change	Monitor anticonvulsant level and adjust dose accordingly
PIs without RTV			
ATV, FPV	Carbamazepine	May ↓ PI levels substantially	Consider alternative anticonvulsant or monitor anticonvulsant level and assess virologic response.

(continued)

Table G.1 (continued)

PI	Other drug	Effect	Comments
LPV/r	Lamotrigine	Lamotrigine AUC ↓ 50% LPV: no significant change	Titrate lamotrigine dose to effect. A similar interaction is possible with other RTV-boosted PIs
All PIs	Phenobarbital	May ↓ PI levels substantially	Consider alternative anticonvulsant or monitor levels of both drugs and assess virologic response. **Do not coadminister with LPV/r once daily**
RTV-boosted PIs			
ATV/r, DRV/r, SQV/r, TPV/r	Phenytoin	↓ Phenytoin possible ↓ PI possible	Consider alternative anticonvulsant or monitor levels of both drugs and assess virologic response
FPV/r		Phenytoin AUC ↓ 22% APV AUC ↑ 20%	Monitor phenytoin level and adjust dose accordingly. No change in FPV/r dose recommended
LPV/r		Phenytoin AUC ↓ 31% LPV/r phenytoin AUC ↓ 33%	Consider alternative anticonvulsant or monitor levels of both drugs and assess virologic response. **Do not coadminister with LPV/r once daily**
PIs without RTV			
ATV, FPV	Phenytoin	May ↓ PI levels substantially	Consider alternative anticonvulsant; RTV boosting for ATV and FPV; and/or monitoring PI level. Monitor anticonvulsant level and virologic response.
LPV/r	Valproic acid (VPA)	↓ VPA possible LPV AUC ↑ 75%	Monitor VPA levels and response. Monitor for LPV-related toxicities
Antidepressants			
LPV/r	Bupropion	Bupropion AUC ↓ 57%	Titrate bupropion dose based on clinical response
TPV/r		Bupropion AUC ↓ 46%	
DRV/r	Paroxetine	Paroxetine ↓ AUC 39%	Titrate paroxetine dose based on clinical response
FPV/r		Paroxetine ↓ AUC 58%	
DRV/r	Sertraline	Sertraline AUC ↓ 49%	Titrate sertraline dose based on clinical response
ATV ± RTV, DRV/r, FPV ± RTV, LPV/r, TPV/r	Trazodone	RTV 200 mg BID (for 2 days) ↑ Trazodone AUC 240%	Use lowest dose of trazodone and monitor for CNS and cardiovascular adverse effects
SQV/r		↑ Trazodone expected	**Contraindicated. Do not coadminister**

(continued)

Table G.1 (continued)

PI	Other drug	Effect	Comments
All RTV-boosted PIs	Tricyclic antidepressants (TCAs) (amitriptyline, desipramine, imipramine, nortriptyline)	↑ TCA expected	Use lowest possible TCA dose and titrate based on clinical assessment and/or drug levels
Antifungals			
RTV-boosted PIs			
ATV/r	Fluconazole	No significant effect	
SQV/r		No data with RTV boosting SQV (1,200 mg TID) AUC ↑ 50%	
TPV/r		TPV AUC ↑ 50%	Fluconazole >200 mg daily is not recommended. If high-dose fluconazole is indicated, consider alternative PI or another class of ARV drug
RTV-boosted PIs			
ATV/r, DRV/r, FPV/r, TPV/r	Itraconazole	↑ Itraconazole possible ↑ PI possible	Consider monitoring itraconazole level to guide dosage adjustments. High doses (>200 mg/day) are not recommended unless dosing is guided by drug levels
LPV/r		↑ Itraconazole	Consider not exceeding 200 mg itraconazole daily or monitor itraconazole level
SQV/r		Bidirectional interaction has been observed	Dose not established, but decreased itraconazole dosage may be warranted. Consider monitoring itraconazole level
PIs without RTV			
ATV, FPV	Itraconazole	↑ Itraconazole possible ↑ PI possible	Consider monitoring itraconazole level to guide dosage adjustments
ATV/r	Posaconazole	ATV AUC ↑ 146%	Monitor for adverse effects of ATV
ATV		ATV AUC ↑ 268%	Monitor for adverse effects of ATV
RTV-boosted PIs			
ATV/r, DRV/r, FPV/r, LPV/r, SQV/r, TPV/r	Voriconazole	RTV 400 mg BID ↓ voriconazole AUC 82% RTV 100 mg BID ↓ voriconazole AUC 39%	**Do not coadminister** voriconazole and RTV unless benefit outweighs risk. If administered, consider monitoring voriconazole level

(continued)

Table G.1 (continued)

PI	Other drug	Effect	Comments
PIs without RTV			
ATV, FPV	Voriconazole	↑ Voriconazole possible ↑ PI possible	Monitor for toxicities
Anti-mycobacterials			
ATV ± RTV	Clarithromycin	Clarithromycin AUC ↑ 94%	May cause QTc prolongation. Reduce clarithromycin dose by 50%. Consider alternative therapy
DRV/r, FPV/r LPV/r, SQV/r, TPV/r		DRV/r ↑ clarithromycin AUC 57% FPV/r ↑ clarithromycin possible LPV/r ↑ clarithromycin expected RTV 500 mg BID ↑ clarithromycin 77% SQV unboosted ↑ clarithromycin 45% TPV/r ↑ clarithromycin 19% and ↓ active metabolite 97% Clarithromycin ↑ unboosted SQV 177% Clarithromycin ↑ TPV 66%	Monitor for clarithromycin-related toxicities Reduce clarithromycin dose by 50% in patients with CrCl 30–60 ml/min Reduce clarithromycin dose by 75% in patients with CrCl < 30 ml/min
FPV		APV AUC ↑ 18%	No dose adjustment
RTV-boosted PIs			
ATV ± RTV	Rifabutin	Rifabutin (150 mg once daily) AUC ↑ 110% and metabolite	Rifabutin 150 mg every other day or three times a week. Some experts recommend rifabutin 150 mg daily or 300 mg three times a week. Monitor for antimycobacterial activity Therapeutic drug monitoring for rifabutin is recommended. Rifabutin 150 mg three times a week in combination with PPV/r has resulted in inadequate rifabutin levels and has led to acquired rifamycin resistance in patients with HIV-associated TB Pharmacokinetic data reported in this table are results from healthy volunteer studies

(continued)

Table G.1 (continued)

PI	Other drug	Effect	Comments
		AUC ↑ 2,101% compared with rifabutin 300 mg daily alone	
DRV/r		Rifabutin (150 mg every other day) and metabolite AUC ↑ 55% compared with rifabutin 300 mg once daily alone	
FPV/r		Rifabutin (150 mg every other day) and metabolite AUC ↑ 64% compared with rifabutin 300 mg once daily alone	
LPV/r		Rifabutin (150 mg once daily) and metabolite AUC ↑ 473% compared with rifabutin 300 mg daily alone	
SQV/r		↑ Rifabutin with unboosted SQV	
TPV/r		Rifabutin (150 mg × 1 dose) and metabolite AUC ↑ 333%	
PIs without RTV			
FPV	Rifabutin	↑ Rifabutin AUC expected	Rifabutin 150 mg daily or 300 mg three times a week
All PIs	Rifampin	↓ PI >75% approximately	**Do not coadminister rifampin and PIs.** Additional RTV does not overcome this interaction and increases hepatotoxicity
Benzodiazepines			
All PIs	Alprazolam	↑ Benzodiazepine possible RTV 200 mg BID for 2 days	Consider alternative benzodiazepines such as lorazepam, oxazepam, or temazepam
	Diazepam	↑ Alprazolam half-life 222% and AUC 248%	
	Lorazepam Oxazepam Temazepam	No data	Metabolism of these benzodiazepines via non-CYP450 pathways decreases interaction potential compared with other benzodiazepines

(continued)

Table G.1 (continued)

PI	Other drug	Effect	Comments
	Midazolam	↑ Midazolam expected SQV/r ↑ midazolam (oral) AUC 1,144% and C_{max} 327%	**Do not coadminister oral midazolam and PIs.** Parenteral midazolam can be used with caution as a single dose and can be given in a monitored situation for procedural sedation
	Triazolam	↑ Triazolam expect-edRTV 200 mg BID ↑ triazolam half-life 1,200% and AUC 2,000%	Do not coadminister triazolam and PIs
Cardiac medications			
All PIs	Bosentan	LPV/r ↑ bosentan 48-fold (day 4) and 5-fold (day 10) ↓ ATV expected	Do not coadminister bosentan and ATV without RTV In patients on a PI (other than unboosted ATV)≥10 days: Start bosentan at 62.5 mg once daily or every other day In patients on bosentan who require a PI (other than unboosted ATV): stop bosentan ≥36 h prior to PI initiation and restart 10 days after PI initiation at 62.5 mg once daily or every other day
RTV, SQV/r	Digoxin	RTV 200 mg BID ↑ digoxin AUC 29% and half-life 43% SQV/r ↑ digoxin AUC 49%	Use with caution. Monitor levels. Digoxin dose may need to be decreased
All PIs	Dihydro-pyridine Calcium channel blockers (CCBs)	↑ Dihydropyridine possible	Use with caution. Titrate CCB dose and monitor closely. ECG monitoring is recommended when used with ATV
ATV ± RTV	Diltiazem	Diltiazem AUC ↑ 125%	Decrease diltiazem dose by 50%. ECG monitoring is recommended
DRV/r, FPV ± RTV, LPV/r, SQV/r, TPV/r		↑ Diltiazem possible	Use with caution. Adjust diltiazem according to clinical response and toxicities
Corticosteroids			
All PIs	Dexametha-sone	↓ PI levels possible	Use systemic dexamethasone with caution or consider alternative corticosteroid for long-term use

(continued)

Table G.1 (continued)

PI	Other drug	Effect	Comments
All RTV-boosted PIs	Fluticasone (inhaled or intranasal)	RTV 100 mg BID ↑ fluticasone AUC 350-fold and ↑ C_{max} 25-fold	Coadministration can result in adrenal insufficiency, including Cushing's syndrome. **Do not coadminister unless potential benefit outweighs risk of systemic corticosteroid adverse effects**
LPV/r	Prednisone	↑ Prednisone AUC 31%	No dosage adjustment necessary
Herbal products			
All PIs	St. John's wort	↓ PI expected	**Do not coadminister**
Hormonal contraceptives			
RTV-boosted PIs			
ATV/r	Hormonal contraceptives	↓ Ethinyl estradiol ↑ Norgestimate	Oral contraceptive should contain at least 35 mcg of ethinyl estradiol. Oral contraceptives containing progestins other than norethindrone or norgestimate have not been studied
DRV/r		Ethinyl estradiol AUC ↓ 44% Norethindrone AUC ↓ 14%	Use alternative or additional method
FPV/r		Ethinyl estradiol AUC ↓ 37% Norethindrone AUC ↓ 34%	Use alternative or additional method
LPV/r		Ethinyl estradiol AUC ↓ 42% Norethindrone AUC ↓ 17%	Use alternative or additional method
SQV/r		↓ Ethinyl estradiol	Use alternative or additional method
TPV/r		Ethinyl estradiol AUC ↓ 48% Norethindrone : no significant change	Use alternative or additional method
PIs without RTV			
ATV	Hormonal contraceptives	Ethinyl estradiol AUC ↑ 48% Norethindrone AUC ↑ 110%	Oral contraceptive should contain no more than 30 mcg of ethinyl estradiol or use alternate method. Oral contraceptives containing less than 25 mcg of ethinyl estradiol or progestins other than norethindrome or norgestimate have not been studied
FPV		With APV: ↑ ethinyl estradiol and ↑ norethindrone; ↓ APV 20%	Use alternative method

(continued)

Table G.1 (continued)

PI	Other drug	Effect	Comments
HMG-CoA reductase inhibitors			
All PIs	Atorvastatin	DRV/r + atorvastatin 10 mg similar to atorvastatin 40 mg alone FPV ± RTV ↑ atorvastatin AUC 130–153% LPV/r ↑ atorvastatin AUC 488% SQV/r ↑ atorvastatin AUC 79% TPV/r ↑ atorvastatin AUC 836%	Use lowest possible starting dose with careful monitoring for toxicities or consider other HMG-CoA reductase inhibitors with less potential for interaction
	Lovastatin	Significant ↑ lovastatin expected	**Contraindicated. Do not coadminister**
ATV	Pitavastatin	Pitavastatin AUC ↑31%; C_{max} ↑60% ATV: no significant effect	No dosage adjustment needed for ATV without RTV
All RTV-boosted PIs		↑ Pitavastatin possible	**Do not coadminister** due to possible increase in pitavastatin concentration and increased risk of rhabdomyolysis
DRV/r	Pravastatin	Pravastatin AUC ↑ 81%	Use lowest possible starting dose with careful monitoring
LPV/r		Pravastatin AUC ↑ 33%	No dose adjustment necessary
SQV/r		Pravastatin AUC ↓ 47–50%	No dose adjustment necessary
ATV/r	Rosuvastatin	↑ Rosuvastatin AUC ↑ 213% and C_{max} ↑ 600%	Use lowest possible starting dose with careful monitoring or consider other HMG-CoA reductase inhibitors with less potential for interaction
DRV/r, FPV ± RTV, SQV/r		↑ Rosuvastatin possible	
LPV/r		Rosuvastatin AUC ↑ 108% and C_{max} ↑366%	
TPV/r		Rosuvastatin AUC ↑ 26% and C_{max} ↑ 123%	
All PIs	Simvastatin	Significant ↑ simvastatin level; SQV/r 400 mg/400 mg BID ↑ Simvastatin AUC 3,059%	**Contraindicated. Do not coadminister**

(continued)

Table G.1 (continued)

PI	Other drug	Effect	Comments
Narcotics/treatment for opioid dependence			
ATV	Buprenorphine	Buprenorphine AUC ↑ 93%	**Do not coadminister buprenorphine with unboosted ATV**
		Norbuprenorphine AUC ↑ 76%	Norbuprenorphine is an active metabolite of buprenorphine
		↓ ATV possible	
ATV/r		Buprenorphine AUC ↑ 66%	Monitor for sedation. Buprenorphine dose reduction may be necessary
		Norbuprenorphine AUC ↑ 105%	Norbuprenorphine is an active metabolite of buprenorphine
DRV/r		Buprenorphine; no significant effect norbuprenorphine AUC ↑ 46% and C_{min} ↑ 71%	No dose adjustment necessary. Clinical monitoring is recommended Norbuprenorphine is an active metabolite of buprenorphine
LPV/r		No significant effect	No dose adjustment necessary
TPV/r		Buprenorphine; no significant effect norbuprenorphine AUC, C_{max} and C_{min} ↓ 80%	Consider monitoring TPV level
		TPV C_{min} ↓ 19–40%	Norbuprenorphine is an active metabolite of buprenorphine
RTV-boosted PIs			
ATV/r, DRV/r, FPV/r, LPV/r, SQV/r, TPV/r	Methadone	ATV/r, DRV/r, FPV/r ↓ R-methadone AUC 16–18%; LPV/r ↓ methadone AUC 26–53% SQV/r 1,000/100 mg BID ↓ R-methadone AUC 19% TPV/r ↓ R-methadone AUC 48%	Opioid withdrawal unlikely but may occur. No adjustment in methadone usually required but monitor for opioid withdrawal and increase methadone dose as clinically indicated (R-methadone is the active form of methadone)
PIs without RTV			
ATV	Methadone	No significant effect	No dosage adjustment necessary
FPV		No data with unboosted FPV APV ↓ R-methadone C_{min} 21%, AUC no significant change	Monitor and titrate methadone as clinically indicated. The interaction with FPV presumed to be similar (R-methadone is the active form of methadone)

(continued)

Table G.1 (continued)

PI	Other drug	Effect	Comments
Phosphodiesterase type 5 (PDE5) inhibitors			
All PIs	Sildenafil	DRV/r+sildenafil 25 mg similar to sildenafil 100 mg alone RTV 500 mg BID ↑ sildenafil AUC 1,000% SQV unboosted ↑ sildenafil AUC 210%	For treatment of erectile dysfunction Start with sildenafil 25 mg every 48 h and monitor for adverse effects of sildenafil For treatment of pulmonary arterial hypertension **Contraindicated**
All PIs	Tadalafil	RTV 200 mg BID ↑ tadalafil AUC 124%; TPV/r (1st dose) ↑ tadalafil AUC 133%; TPV/r steady state; no significant effect	For treatment of erectile dysfunction Start with tadalafil 5 mg dose and do not exceed a single dose of 10 mg every 72 h. Monitor for adverse effects of tadalafil For treatment of pulmonary arterial hypertension *In patients on a PI >7 days:* Start with tadalafil 20 mg once daily and increase to 40 mg once daily based on tolerability *In patients on tadalafil who require a PI:* Stop tadalafil ≥24 h prior to PI initiation, restart 7 days after PI initiation at 20 mg once daily, and increase to 40 mg once daily based on tolerability
All PIs	Vardenafil	RTV 600 mg BID ↑ vardenafil AUC 49-fold	Start with vardenafil 2.5 mg every 72 h and monitor for adverse effects of vardenafil
Miscellaneous interactions			
Colchicine	All PIs	All PIs	For treatment of gout flares Colchicine 0.6 mg × 1 dose, followed by 0.3 mg 1 h later. Do not repeat dose for at least 3 days. *With FPV without RTV:* 1.2 mg × 1 dose and no repeat dose for at least 3 days For prophylaxis of gout flares Colchicine 0.3 mg once daily or every other day. *With FPV without RTV:* colchicine 0.3 mg BID or 0.6 mg once daily or 0.3 mg once daily For treatment of familial Mediterranean fever Do not exceed colchicine 0.6 mg once daily or 0.3 mg BID. *With FPV without RTV:* Do not exceed 1.2 mg once daily or 0.6 mg BID Do not coadminister in patients with hepatic or renal impairment

(continued)

Table G.1 (continued)

PI	Other drug	Effect	Comments
Salmeterol	All PIs	↑ Salmeterol possible	**Do not coadminister** because of potential increased risk of salmeterol-associated cardiovascular events, including QT prolongation, palpitations, and sinus tachycardia
Atovaquone/ proguanil	ATV/r	ATV/r ↓ atovaquone AUC 46% and ↓ proguanil AUC 41%	No dosage recommendation, consider alternative drug for malaria prophylaxis, if possible
	LPV/r	LPV/r ↓ atovaquone AUC 74% and ↓ proguanil AUC 38%	

Acronyms: APV amprenavir, *ARV* antiretroviral, *ATV* atazanavir, *ATV/r* atazanavir + ritonavir, *AUC* area under the curve, *BID* twice daily, *CCB* calcium channel blocker, C_{max} maximum plasma concentration, C_{min} minimum plasma concentration, *CNS* central nervous system, *CrCl* creatinine clearance, *CYP* cytochrome P, *DRV/r* darunavir + ritonavir, *ECG* electrocardiogram, *FDA* Food and Drug Administration, *FPV* fosamprenavir (FPV is a prodrug of APV), *FPV/r* fosamprenavir + ritonavir, *IDV* indinavir, *IDV/r* indinavir + ritonavir, *INR* international normalized ratio, *LPV* lopinavir, *LPV/r* lopinavir + ritonavir, *NFV* nelfinavir, *PDE5* phosphodiesterase type 5, *PI* protease inhibitor, *PPI* proton pump inhibitor, *RTV* ritonavir, *SQV* saquinavir, *SQV/r* saquinavir + ritonavir, *TB* tuberculosis, *TCA* tricyclic antidepressant, *TID* three times a day, *TPV* tipranavir, *TPV/r* tipranavir + ritonavir, *VPA* valproic acid

Table G.2 Drug interactions between NNRTIs and other drugs

NNRTI	Other	Effect	Comments
Anticoagulants/antiplatelets			
EPV, NVP	Warfarin	↑ or ↓ Warfarin possible	Monitor INR and adjust warfarin dose accordingly
ETR		↑ Warfarin possible	Monitor INR and adult warfarin dose accordingly
ETR	Clopidogrel	↓ Activation of clopidogrel possible	ETR may prevent metabolism of clopidogrel (inactive) to its active metabolite. Avoid coadministration, if possible
Anticonvulsants			
EFV	Carbamazepine Phenobarbital Phenytoin	Carbamazepine + EFV Carbamazepine AUC ↓ 27% and EFV AUC ↓36% Phenytoin + EFV: ↓ EFV and ↓ phenytoin possible	Monitor anticonvulsant and EFV levels or, if possible, use alternative anticonvulsant
ETR		↓ Anticonvulsant and ETR possible	**Do not coadminister.** Consider alternative anticonvulsants
NVP		↓ Anticonvulsant and NVP possible	Monitor anticonvulsant and NVP levels and virologic responses
Antidepressants			
EFV	Bupropion	Bupropion AUC ↓ 55%	Titrate bupropion dose based on clinical response
ETR	Paroxetine	No significant effect	No dosage adjustment necessary
EFV	Sertraline	Sertraline AUC ↓ 39%	Titrate sertraline dose based on clinical response
Antifungals			
EFV	Fluconazole	No significant effect	
ETR		ETR AUC ↑ 86%	No dosage adjustment. Use with caution
NVP		NVP AUC ↑ 110%	Increased risk of hepatotoxicity possible with this combination. Monitor NVP toxicity or use alternative antiretroviral agent
EFV	Itraconazole	Itraconazole and OH-itraconazole AUC, C_{max} and C_{min} ↓ 35–44%	Dose adjustments for itraconazole may be necessary. Monitor itraconazole level and antifungal response
ETR		↓ Itraconazole possible ↑ ETR possible	Dose adjustments for itraconazole may be necessary. Monitor itraconazole level and antifungal response

(continued)

Table G.2 (continued)

NNRTI	Other	Effect	Comments
NVP		↓ Itraconazole possible ↑ NVP possible	Consider monitoring NNRTI and itraconazole levels and antifungal response
EFV	Posaconazole	Posaconazole AUC ↓ 50% ↔ EFV	Consider alternative antifungal if possible or consider monitoring posaconazole level if available
ETR		↑ ETR possible	No dosage adjustment necessary
EFV	Voriconazole	Voriconazole AUC ↓ 77% EFV AUC ↑ 44%	**Contraindicated at standard doses.** Dose: Voriconazole 400 mg BID, EFV 300 mg daily
ETR		Voriconazole AUC ↑ 14% ETR AUC ↑ 36%	No dosage adjustment; use with caution Consider monitoring voriconazole level
NVP		↓ Voriconazole possible ↑ NVP possible	Monitor for toxicity and antifungal response and/or voriconazole level
Antimycobacterials			
EFV	Clarithromycin	Clarithromycin AUC ↓ 39%	Monitor for efficacy or consider alternative agent, such as azithromycin, for MAC prophylaxis and treatment
ETR		Clarithromycin AUC ↓ 39% OH-clarithromycin AUC ↑ 21% ETR AUC ↑ 42%	Monitor for efficacy or use alternative agent, such as azithromycin, for MAC prophylaxis and treatment
NVP	Clarithromycin	Clarithromycin AUC ↓ 31% OH-clarithromycin AUC ↑ 42%	Monitor for efficacy or use alternative agent, such as azithromycin, for MAC prophylaxis and treatment
EFV	Rifabutin	Rifabutin ↓ 38%	Dose: rifabutin 450–600 mg once daily or 600 mg three times a week if EFV is not coadministered with a PI
ETR		Rifabutin and metabolite AUC ↓ 17% ETR AUC ↓ 37%	**If ETR is used with an RTV-boosted PI, rifabutin should not be coadministered** Dose: rifabutin 300 mg once daily **if** ETR is not coadministered with and RTV-boosted PI
NVP		Rifabutin and metabolite AUC ↑ 17% and metabolite AUC ↑ 24% NVP C_{min} ↓16%	No dosage adjustment necessary. Use with caution
EFV	Rifampin	EFV AUC ↓ 26%	Maintain EFV dose at 600 mg once daily and monitor for virologic response. Some clinicians suggest EFV 800 mg dose in patients >60 kg

(continued)

Table G.2 (continued)

NNRTI	Other	Effect	Comments
ETR		Significant ↓ ETR possible	**Do not coadminister**
NVP		NVP ↓ 20–58%	**Do not coadminister**
Benzodiazepines			
EFV, ETR, NVP	Alprazolam	No data	Monitor for therapeutic efficacy of alprazolam
ETR	Diazepam	↑ Diazepam possible	Decreased dose of diazepam may be necessary
EFV	Lorazepam	Lorazepam C_{max} ↑ 16%, AUC no significant effect	No dosage adjustment necessary
EFV	Midazolam	Significant ↑ in midazolam expected	**Do not coadminister with oral midazolam** Parenteral midazolam can be used with caution as a single dose and can be given in a monitored situation for procedural sedation
EFV	Triazolam	Significant ↑ triazolam expected	**Do not coadminister**
Cardiac medications			
EFV, NVP	Dihydropyridine calcium channel blockers (CCBs)	↓ CCBs possible	Titrate CCB dose based on clinical response
	Diltiazem		
EFV		Diltiazem AUC ↓ 69%	Titrate diltiazem dose based on clinical response
NVP		↓ Diltiazem possible	
Corticosteroids			
ETR	Dexamethasone	↓ ETR possible	Use systemic dexamethasone with caution or consider alternative corticosteroid for long-term use
Herbal products			
EFV, ETR, NVP	St. John's wort	↓ NNRTI	**Do not coadminister**
Hormonal contraceptives			
EFV	Hormonal contraceptives	Ethinyl estradiol ↔ levonorgestrel AUC ↓ 83% Norelgestromin AUC ↓ 64%	Use alternative or additional methods. Norelgestromin and levonorgestrel are active metabolites of norgestimate
ETR		Ethinyl estradiol AUC ↑ 22% Norethindrone: no significant effect	No dosage adjustment necessary

(continued)

Table G.2 (continued)

NNRTI	Other	Effect	Comments
NVP		Ethinyl estradiol AUC ↓20% Norethindrone AUC ↓19%	Use alternative or additional methods
		Depomedroxyprogesterone acetate: no significant change	No dosage adjustment necessary
EFV	Levonorgestrel	Levonorgestrel AUC ↓ 58%	Effectiveness of emergency postcoital contraception may be diminished
HMG-CoA reductase inhibitors			
EFV, ETR, NVP	Atorvastatin	Atorvastatin AUC ↓ 32–43% with EFV, ETR	Adjust atorvastatin according to lipid responses, not to exceed the maximum recommended dose
ETR	Fluvastatin	↑ Fluvastatin possible	Dose adjustments for fluvastatin may be necessary
EFV	Lovastatin Simvastatin	Simvastatin AUC ↓68%	Adjust simvastatin dose according to lipid responses, not to exceed the maximum recommended dose. If used with RTV-boosted PI, simvastatin and lovastatin should be avoided
ETR, NVP		↓ Lovastatin possible ↓ Simvastatin	Adjust lovastatin or simvastatin dose according to lipid responses, not to exceed the maximum recommended dose. If used with RTV-boosted PI, simvastatin and lovastatin should be avoided
EFV, ETR, NVP	Pitavastatin	No data	No dosage recommendation
EFV	Pravastatin Rosuvastatin	Pravastatin AUC ↓ 44% Rosuvastatin: no data	Adjust statin dose according to lipid responses, not to exceed the maximum recommended dose
ETR		No significant effect expected with either pravastatin or rosuvastatin	No dosage adjustment necessary
Narcotics/treatment for opioid dependence			
EFV	Buprenorphine	Buprenorphine AUC ↓ 50% Norbuprenorphine AUC ↓ 71%	No withdrawal symptoms reported. No dosage adjustment recommended, but monitor for withdrawal symptoms
NVP		No significant effect	No dosage adjustment necessary
EFV	Methadone	Methadone AUC ↓ 52%	Opioid withdrawal common; increased methadone dose often necessary
ETR		No significant effect	No dosage adjustment necessary

(continued)

Table G.2 (continued)

NNRTI	Other	Effect	Comments
NVP		Methadone AUC ↓ 41% NVP: no significant effect	Opioid withdrawal common; increased methadone dose often necessary
Phosphodiesterase type 5 (PDE5) inhibitors			
ETR	Sildenafil	Sildenafil AUC ↓ 57%	May need to increase sildenafil dose based on clinical effect
ETR	Tadalafil	↓ Tadalafil possible	May need to increase tadalafil dose based on clinical effect
ETR	Vardenafil	↓ Vardenafil possible	May need to increase vardenafil dose based on clinical effect
Miscellaneous interactions			
EFV	Atovaquone/ proguanil	↓ Atovaquone AUC 75% ↓ Proguanil AUC 43%	No dosage recommendation. Consider alternative drug for malaria prophylaxis, if possible

Acronyms: ARV antiretroviral, *AUC* area under the curve, *CCB* calcium channel blocker, C_{max} maximum plasma concentration, C_{min} minimum plasma concentration, *DLV* delavirdine, *EFV* efavirenz, *ETR* etravirine, *FDA* Food and Drug Administration, *INR* international normalized ratio, *MAC Mycobacterium avium* complex, *NNRTI* non-nucleoside reverse transcriptase inhibitor, *NVP* nevirapine, *OH-clarithromycin* active metabolite of clarithromycin, *PDE5* phosphodiesterase type 5, *PI* protease inhibitor

Table G.3 Drug interactions between NRTIs and other drugs (including ARV agents)

NRTI	Other	Effect	Comment
Antivirals			
TDF	Ganciclovir Valgan-ciclovir	No data	Serum concentrations of these drugs and/or TDF may be increased. Monitor for dose-related toxicities
ZDV		No significant pharmacokinetic effects	Potential increase in hematologic toxicities
ddI	Ribavirin	↑ Intracellular ddI	**Contraindicated. Do not coadminister.** Fatal hepatic failure and other ddI-related toxicities have been reported with coadministration
ZDV		Ribavirin inhibits phosphorylation of ZDV	Avoid coadministration if possible or closely monitor virologic response and hematologic toxicities
Integrase inhibitor			
TDF	RAL	RAL AUC ↑ 49%, C_{max} ↑ 64%	No dosage adjustment necessary
Narcotics/treatment for opioid dependence			
3TC, ddI, TDF, ZDV	Buprenor-phine	No significant effect	No dosage adjustment necessary

(continued)

Table G.3 (continued)

NRTI	Other	Effect	Comment
ABC	Methadone	Methadone clearance ↑ 22%	No dosage adjustment necessary
d4T		D4T AUC ↓ 23% and C_{max} ↓ 44%	No dosage adjustment necessary
ZDV		ZDV AUC ↑ 29–43%	Monitor for ZDV-related adverse effects
NRTIs			
d4T	ddI	No significant PK interaction	**Avoid coadministration.** Additive toxicities of peripheral neuropathy, lactic acidosis, and pancreatitis seen with this combination
TDF		ddI-EC AUC and C_{max} ↑ 48–60%	**Avoid coadministration**
Other			
ddI	Allopurinol	ddI AUC ↑ 113% ddI AUC ↑ 312% with renal impairment	**Contraindicated. Do not coadminister.** Potential for increased ddI-associated toxicities
PIs			
ddI	ATV	With ddI-EC + ATV (with food): ddI AUC ↓ 34%; ATV no change	Administer ATV with food 2 h before or 1 h after didanosine
TDF		ATV AUC ↓ 25% and C_{min} ↓ 23–40% (higher C_{min} with RTV than without) TDF AUC ↑ 24–37%	Dose: ATV/r 300/100 mg daily coadministered with TDF 300 mg daily. Avoid concomitant use without RTV. If using TDF and H_2 receptor antagonist in ART-experienced patients, use ATV/r 400 mg/100 mg daily Monitor for TDF-associated toxicity
ZDV		ZDV C_{min} ↓ 30%, no change in AUC	Clinical significance unknown
TDF	DRV/r	TDF AUC ↑ 22%, C_{max} ↑ 24% and C_{min} ↑ 37%	Clinical significance unknown. Monitor for TDF toxicity
TDF	LPV/r	LPV/r AUC ↓ 15% TDF AUC ↑ 34%	Clinical significance unknown. Monitor for TDF toxicity
ABC	TPV/r	ABC ↓ 35–44% with TPV/r 1,250/100 mg BID	Appropriate doses for this combination have not been established
ddI		ddI-EC ↓ 10% and TPV C_{min} ↓ 34% with TPV/r 1,250/100 mg BID	Separate doses by at least 2 h
ZDV		ZDV AUC ↓ 31–43% and C_{max} ↓ 46–51% with TPV/r 1,250/100 mg BID	Appropriate doses for this combination have not been established

Acronyms: 3TC lamivudine, *ABC* abacavir, *ARV* antiretroviral, *ATV* atazanavir, *AUC* area under the curve, *BID* twice daily, C_{max} maximum plasma concentration, C_{min} minimum plasma concentration, *d4T* stavudine, *ddI* didanosine, *DRV/r* darunavir/ritonavir, *EC* enteric coated, *LPV/r* lopinavir/ritonavir, *NRTI* nucleoside reverse transcriptase inhibitor, *PI* protease inhibitor, *PK* pharmacokinetic, *RAL* raltegravir, *TDF* tenofovir, *TPV/r* tipranavir/ritonavir, *ZDV* zidovudine

Table G.4 Drug interactions between CCR5 antagonist and other drugs

CCR5 antagonist	Other	Effect	Comments
Anticonvulsants			
MVC	Carbamazepine Phenobarbital Phenytoin	↓ MVC possible	If used without a strong CYP3A inhibitor, use MVC 600 mg BID or an alternative antiepileptic agent
Antifungal			
MVC	Itraconazole	↑ MVC possible	Done MVC 150 mg BID
MVC	Ketoconazole	MVC AUC ↑ 400%	Dose: MVC 150 mg BID
MVC	Voriconazole	↑ MVC possible	Consider dose reduction to MVC 150 mg BID
Antimycobacterials			
MVC	Clarithromycin	↑ MVC possible	Dose: MVC 150 mg BID
MVC	Rifabutin	↓ MVC possible	If used without a strong CYP3A inducer or inhibitor, use MVC 300 mg BID If used with a strong CYP3A inhibitor, use MVC 150 mg BID
MVC	Rifampin	MVC AUC ↓ 64%	Coadministration is not recommended. If coadministration is necessary use MVC 600 mg BID. If coadministered with a strong CYP3A inhibitor, use MVC 300 mg BID
Herbal products			
MVC	St. John's wort	↓ MVC possible	Coadministration is not recommended
Hormonal contraceptives			
MVC	Hormonal contraceptives	No significant effect on ethinyl estradiol or levonorgestrel	Safe to use in combination
Narcotics/treatment for opioid dependence			
MVC	Methadone	No data	

Acronyms: *ARV* antiretroviral, *AUC* areas under the curve, *BID* twice daily, *CYP* cytochrome P, *MVC* maraviroc

Table G.5 Drug interactions between integrase inhibitor and other drugs

Integrase inhibitor	Drug	Effect	Comment
Acid reducers			
RAL	Omeprazole	RAL AUC ↑ 212%, C_{max} ↑ 315%, and C_{min} ↑ 46%	No dosage adjustment recommended
Antimycobacterials			
RAL	Rifabutin	RAL AUC ↑ 19%, C_{max} ↑ 39%, and C_{min} ↓ 20%	No dosage adjustment recommended
RAL	Rifampin	RAL AUC ↓ 40% and C_{min} ↓ 61% with RAL 400 mg	
		Rifampin with RAL 800 mg BID compared with RAL 400 mg BID alone; RAL AUC ↑ 27% and C_{min} ↓ 53%	Dose: RAL 800 mg BID. Monitor closely for virologic response
Hormonal contraceptives			
RAL	Hormonal contraceptives	No clinically significant effect	Safe to use in combination
Narcotics/treatment for opioid dependence			
RAL	Methadone	No significant effect	No dosage adjustment necessary

Acronyms: AUC area under the curve, *BID* twice daily, C_{max} maximum plasma concentration, C_{min} minimum plasma concentration, *RAL* raltegravir

Table G.6 Interactions among PIs

Drug affected	ATV	FPV	LPV/r	RTV	SQV	TPV
DRV	Dose: ATV 300 mg once daily + DRV 600 mg BID + RTV 100 mg BID	No data	**Should not be coadministered because doses are not established**	Dose: (DRV 600 mg + RTV 100 mg) BID or (DRV 800 mg + RTV 100 mg) once daily	**Should not be coadministered because doses are not established**	No data
FPV	Dose: Insufficient data	–	**Should not be coadministered because doses are not established**	Dose: (FPV 1,400 mg + RTV 100 mg or 200 mg) once daily; or (FPV 700 mg + RTV 100 mg) BID	Dose: insufficient data	**Should not be coadministered because doses are not established**
LPV/r	Dose: ATV 300 mg once daily + LPV/r 400/100 mg BID	See LPV/r + FPV cell	–	LPV is coformulated with RTV as Kaletra	See LPV/r + SQV cell	**Should not be coadministered because doses are not established**
RTV	Dose: (ATV 300 mg + RTV 100 mg) once daily	See RTV + FPV cell	LPV is coformulated with RTV as Kaletra	–	Dose: (SQV 1,000 mg + RTV 100 mg) BID	Dose: (TPV 500 mg + RTV 200 mg) BID
SQV	Dose: Insufficient data	Dose: Insufficient data	Dose: SQV 1,000 mg BID + LPV/r 400/100 mg BID	See SQV + RTV cell	–	**Should not be coadministered because doses are not established**

Acronyms: *ATV* atazanavir, *BID* twice daily, *DRV* darunavir, *FDA* Food and Drug Administration, *FPV* fosamprenavir, *IDV* indinavir, *LPV/r* lopinavir/ritonavir, *NFV* nelfinavir, *PI* protease inhibitor, *RTV* ritonavir, *SQV* saquinavir, *TPV* tipranavir

Table G.7 Interactions between NNRTIs, MVC, RAL, and PIs

		EFV	ETR	NVP	MVC	RAL
ATV	PK data	With unboosted ATV ATV: AUC ↓ 74% EPV: no significant change With (ATV 300 mg ± RTV 100 mg) once daily with food. ATV concentrations similar to unboosted ATV without EFV	With unboosted ATV ETR:AUC ↑50% C_{max} ↑ 47% and C_{min} ↑ 58% ATV:AUC ↓ 17% and C_{min} ↓ 47% With (ATV 300 mg ± RTV 100 mg) once daily ETR:AUC, C_{max} and C_{min} ↑ approximately 37% ATV:AUC ↓14% and C_{min} ↓ 38%	With (ATV 300 mg ± RTV 100 mg) once daily ATV: AUC ↓ 42% and C_{min} ↓ 72% NVP: AUC ↑ 25%	With unboosted ATV. MVC AUC ↑ 257% With (ATV 300 mg ± RTV 100 mg) once daily MVC: AUC ↑ 388%	With unboosted ATV RAL: AUC ↑ 72% With (ATV 300 mg ± RTV 100 mg) once daily RAL: AUC ↑ 41%
	Dose	**Do not coadminister with unboosted ATV** In ART-naïve patients (ATV 400 mg + RTV 100 mg) once daily **Do not coadminister in ART-experienced patients**	**Do not coadminister With ATV ± RTV**	**Do not coadminister with ATV ± RTV**	MVC 150 mg BID with ATV ± RTV	Standard

DRV – always use with RTV	PK data	With (DRV 300 mg ± RTV 100 mg) BID DRV: AUC ↓ 13%, C_{min} ↓ 31% EFV: AUC ↑21%	ETR 100 mg BID with (DRV 600 mg + RTV 100 mg) BID DRV: no significant change ETR: AUC ↓ 37%, C_{min} ↓ 49%	With (DRV 400 mg ± RTV 100 mg BID) DRV: AUC ↑ 24% NVP: AUC ↑ 27% and C_{min} ↑ 47%	With (DRV 600 mg ± RTV 100 mg BID) MVC: AUC ↑ 305% With (DRV 600 mg ± RTV 100 mg) BID ± ETR MVC: AUC ↑ 210%	With (DRV 600 mg ± RTV 100 mg BID RAL: AUC ↓ 29% and C_{min} ↑ 38%
	Dose	Clinical significance unknown. Use standard doses and monitor closely. Consider monitoring levels	Standard (ETR 200 mg BID. Despite decreased ETR, safety and efficacy established with this combination in a clinical trial	Standard	MVC 150 mg BID	Standard
EFV	PK data	–	↓ ETR possible	NVP: no significant change. EFV: AUC ↓ 22%	MVC: AUC ↓ 45%	EFV: AUC ↓ 36%
	Dose		**Do not coadminister**	**Do not coadminister**	MVC: 600 mg BID	Standard
ETR	PK data	↓ ETR possible	–	↓ETR possible	MVC: AUC ↓ 53%, C_{max} ↓ 60%	ETR: C_{min} ↓ 17% RAL: C_{min} ↓ 34%
	Dose	**Do not coadminister**		**Do not coadminister**	MVC 600 mg BID	Standard

(continued)

Table G.7 (continued)

		EFV	ETR	NVP	MVC	RAL
FPV	PK data	With (FPV 1,400 mg ± RTV 200 mg once daily APV: C_{min} ↓ 36%	With (FPV 700 mg ± RTV 100 mg BID APV: AUC ↑ 69% C_{min} ↑ 77%	With unboosted FPV 1,400 mg BID APV: AUC ↓ 33% NVP: AUC ↑ 29% With (FPV 1,400 mg ± RTV 100 mg) BID NVP: standard (FPV 700 mg + RTV 100 mg) BID NVP: C_{min} ↑ 19%	Unknown: ↑ MVC possible	No data
	Dose	(FPV 1,400 mg ± RTV 300 mg) once daily; or (FPV 700 mg + RTV 100 mg) BID EFV standard	**Do not coadminister with FPV±RTV**	(FPV 700 mg + RTV 100 mg) BID NVP standard	MVC 150 mg BID	Standard
LPV/r	PK data	With LPV/r tablets 500/125 mg BID↑ + EFV 600 mg LPV levels similar to LPV/r 400/100 mg BID without EFV	With LPV/r tablets ETR: levels ↓ 30–45% (comparable to the decrease with DRV/r LPV levels ↓ 13–20%	With LPV/r capsules LPV: AUC ↓ 27% and C_{min} ↓ 31%	MVC AUC ↑ 295% With LPV/r ± EFV MVC: AUC ↑ 153%	↓ RAL ↔ LPV/r
	Dose	LPV/r tablets 500/125 mg ↕ BID; LPV/r oral solution 533/133 mg BID EFV standard	Standard	LPV/r tablets 500/125 mg BID, LPV/r oral solution 533/133 mg BID NVP standard	MVC 150 mg BID	Standard

NVP	PK Data	NVP: no significant change EFV: AUC ↓ 22%	↓ ETR possible	–	No significant change	No data
	Dose	**Do not coadminister**	**Do not coadminister**		Without PI MVC 300 mg BID With PI (except TPV/r) MVC:150 mg BID	Standard
RAL	PK data	RAL: AUC ↓ 36%	ETR: C_{min} ↑ 17% RAL: C_{min} ↓ 34%	No data	RAL: AUC ↓ 37% MVC: AUC ↓ 21%	–
	Dose	Standard	Standard	No data	Standard	Standard
RTV	PK data	Refer to information for boosted PI	Refer to information for boosted PI	Refer to information for boosted PI	With RTV 100 mg BID MVC AUC ↑ 161%	With RTV 100 mg BID RAL: AUC ↓ 16%
	Dose				MVC 150 mg BID	Standard
SQV – always use with RTV	PK data	With SQV 1,200 mg TID SQV: AUC ↓ 62% EFV: AUC ↓ 12%	With (SQV 1,000 mg ± RTV 100 mg) BID SQV: AUC unchanged ETR: AUC ↓ 33%, C_{min} ↓ 29% Reduced ETR levels similar to reduction with DRV/r	With SQV 600 mg TID SQV: AUC ↓ 38% NVP: no significant change	With (SQV 1,000 mg ± RTV 100 mg) BID MVC: AUC ↑ 877% With (SQV 1,000 mg ± RTV 100 mg) BID ± EFV MVC:AUC ↑ 400%	No data
	Dose	(SQV 1,000 mg + RTV 100 mg) BID	(SQV 1,000 mg + RTV 100 mg) BID	(SQV 1,000 mg + RTV 100 mg) BID	MVC 150 mg BID	Standard

(continued)

Table G.7 (continued)

		EFV	ETR	NVP	MVC	RAL
TPV – always use with RTV	PK data	With TPV 500 mg ± RTV 100 mg) BID TPV: AUC ↓ 31%, C_{min} ↓ 42% EFV: no significant change With (TPV 750 mg ± RTV 200 mg) BID TPV: no significant change EFV: No significant change	With (TPV 500 mg ± RTV 200 mg) BID ETR: AUC ↓ 76%, C_{min} ↓ 82% TPV: AUC ↑ 18%, C_{min} ↑ 24%	With (TPV 250 mg ± RTV 200 mg) and with (TPV 750 mg ± RTV 100 mg) BID NVP: no significant change TPV: no data	With (TPV 500 mg ± RTV 200 mg) BID MVC: no significant change in AUC TPV: no data	With (TPV 500 mg ± RTV 200 mg) BID RAL: AUC ↓ 24%
	Dose	Standard	**Do not coadminister**	Standard	MVC 300 mg BID	Standard

Acronyms: AUC area under the curve, *ATV* atazanavir, *BID* twice daily, C_{max} maximum plasma concentration, C_{min} minimum plasma concentration, *DLV* delaviridine, *DRV* darunavir, *EFV* efavirenz, *ETR* etravirine, *FPV* fosamprenavir, *IDV* indinavir, *LPV/r* lopinavir/ritonavir, *MVC* maraviric, *NFV* nelfinavir, *NNRTI* non-nucleoside reverse transcriptase inhibitor, *NVP* nevirapine, *PI* protease inhibitor, *PK* pharmacokinetic, *RAL* raltegravir, *RTV* ritonavir, *SQV* saquinavir, *TID* three times a day, *TPV* tipranavir

Further Reading

Guidelines for the use of antiretroviral agents in HIV-1 infected adults and adolescents. http://aidsinfo.nih.gov/contentfiles/AdultandAdolescentGL.pdf; accessed on May 12, 2011

Index

CPSIA information can be obtained at www.ICGtesting.com
Printed in the USA
LVOW102145070312
272121LV00007B/131/P